Polarization in Optical Fibers

The Artech House Applied Photonics Series

Series Editors

Brian Culshaw
Alan Rogers

For a listing of recent related Artech House titles,
turn to the back of this book.

Polarization in Optical Fibers

Alan Rogers

ARTECH
HOUSE

BOSTON | LONDON
artechhouse.com

Library of Congress Cataloging-in-Publication Data
A catalog record for this book is available from the U.S. Library of Congress.

British Library Cataloguing in Publication Data
A catalogue record for this book is available from the British Library.

ISBN-13: 978-1-58053-534-2

Cover design by Yekaterina Ratner

The following diagrams are reprinted, with permission, from *Essentials of Opto-Electronics*, by the same author, and published by Chapman and Hall, 1997: Figures 1.1–1.18, 2.1–2.19, 3.1–3.15, 4.1–4.17, 5.1, 5.23–5.26, 6.1, 6.2, 6.4–6.16, 7.4, and 7.11–7.13.

685 Canton Street
Norwood, MA 02062

10 9 8 7 6 5 4 3 2 1

To my wife, Wendy, and my two sons, Daniel and Gareth

Contents

Preface

The use of waveguides for guiding light is well established for a variety of purposes, from high-speed telecommunications, through optical sensing, to the medically important delivery of laser pulses into the human body. For the majority of these uses the requirement is for effective delivery of coded optical power.

However, light is a transverse wave, and its detailed properties as such are described by its polarization behavior; and these properties become especially important whenever there is transverse asymmetry in the medium through which the light is propagating. This helps to characterize the medium and allows control of the light.

Optical fibers can possess transverse asymmetry for a variety of reasons: when the fiber is bent or twisted, when the core cross-section is not a true circle, or when it is subjected to electric or magnetic fields, for example. Sometimes these imposed asymmetries have nuisance value (optical telecommunications is a case in point), but sometimes they can be used to considerable advantage (as in various optical sensors).

In both cases the polarization phenomena need to be thoroughly understood, in order either to mitigate the disadvantages, or to enhance and control the advantages.

This book attempts to provide this understanding at the essential physical level. Inevitably, some mathematics is necessary in order to properly quantify the phenomena, but always the emphasis will be on physical mechanism rather than on detailed mathematical development. Where more detailed mathematics is crucial, it is placed in an appendix.

Optical fibers provide a powerful and, in many ways, elegant medium for the appreciation of, and insight into, the rewarding study of polarization optics and its uses.

I am very grateful to all of my professional colleagues who, by interactions over many years, have, through their sharp intelligences and reliable friendship, enhanced my understanding and enjoyment of the photonics discipline. They have helped considerably to make this book possible.

Finally, I would like to thank my wife, Wendy, for her help and forbearance during the writing of this book.

1

The Wave Theory of Light

1.1 Introduction

Electromagnetic radiation, including light, exhibits both wave and particle properties, and the type of behavior exhibited at any one time depends upon the special circumstances. In this chapter we shall concentrate just on relevant wave properties, for almost all of the polarization phenomena that occur in optical waveguides, including optical fibers, can best be understood within the wave theory.

The essentials of the wave theory of light were discovered and examined in the nineteenth century, before the advent of quantum mechanics (in 1901). The success of the wave theory was remarkable, and it led to a number of important devices, some of which are described in this chapter.

We shall begin by looking at some aspects of the wave theory's crowning glory: Maxwell's equations for the electromagnetic field.

1.2 Electromagnetic Waves

1.2.1 Velocity and Refractive Index

In 1864 James Clerk Maxwell showed conclusively that light waves were electromagnetic in nature, consisting of electric and magnetic fields oscillating orthogonally to each other, and to the direction of the propagation of the wave. He did this by expressing the then-known laws of electromagnetism in such a way as to allow him to derive from them a wave equation (see Appendix A). This wave equation permitted free-space solutions that corresponded to electromagnetic waves with a velocity equal to the known experimental value of the velocity of light. The consequent recognition of light as an electromagnetic phenomenon was probably the single most important advance in the progression of its understanding.

All the important features of light waves follow from a detailed examination of Maxwell's equations (see Appendix A). Taking Cartesian axes *Ox*, *Oy*, *Oz* (Figure 1.1), a typical sinusoidal solution is given by

$$E_x = E_0 \exp[i(\omega t - kz)] \tag{1.1}$$

$$H_y = H_0 \exp[i(\omega t - kz)]$$

which states that the electric field oscillates sinusoidally in the *xz* plane, the magnetic field oscillates in the *yz* plane (i.e., orthogonally to the *E* field) and in phase with the *E* field, and the wave propagates in the *Oz* direction (Figure 1.1). The frequency and wavelength of the wave are given by

$$f = \frac{\omega}{2\pi}$$

$$\lambda = \frac{2\pi}{k}$$

and $f\lambda = \omega/k = c$ where c is the wave velocity. The latter is related to the electromagnetic properties of the medium in which the wave propagates via the relation

$$c = (\epsilon\mu)^{-1/2} \tag{1.2}$$

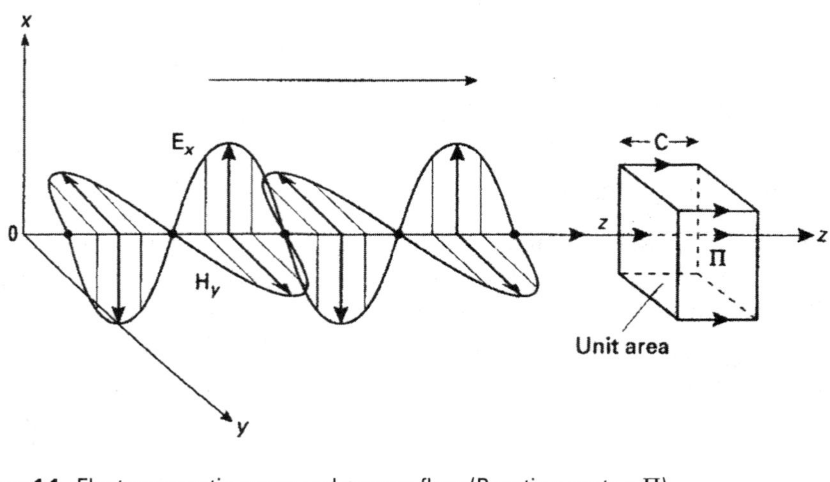

Figure 1.1 Electromagnetic wave and energy flow (Poynting vector: Π).

where ϵ is the electric permittivity of the medium, and μ is its magnetic permeability. The relation (1.2) can also be written in the following form:

$$c = (\epsilon_R \epsilon_0 \mu_R \mu_0)^{-1/2}$$

that is,

$$\epsilon = \epsilon_R \epsilon_0, \ \mu = \mu_R \mu_0$$

where ϵ_R, μ_R are the permittivity and permeability factors for the medium relative to those for free space, ϵ_0, μ_0; ϵ_R is often called the dielectric constant. The electric displacement **D** and the magnetic flux density **B** are defined by the following relations:

$$\mathbf{D} = \boldsymbol{\epsilon} \mathbf{E}$$

$$\mathbf{B} = \boldsymbol{\mu} \mathbf{H}$$

(For reasons of symmetry, **D** is sometimes called the electric flux density.) We can, therefore, also write

$$c = \frac{c_0}{(\epsilon_R \mu_R)^{1/2}} \tag{1.3}$$

where c_0 is the velocity of the electromagnetic wave in free space, and has the (defined) value:

$$2.99792458 \times 10^8 \text{ m.s}^{-1}$$

For most optical media of any importance we have $\mu_R \sim 1$ and $\epsilon_R > 1$. These materials belong to the class known as dielectrics and many are electrical insulators. Thus we may write (1.3) in the form

$$c \approx \frac{c_0}{(\epsilon_R)^{1/2}}$$

and note that $c < c_0$. The ratio c_0/c is, by definition, the refractive index n of the medium, so that

$$n \approx \epsilon_R^{1/2} \tag{1.4}$$

where n is thus the factor by which light travels more slowly in an optical medium than it does in free space. Now ϵ_R is a measure of the ease with which the medium can be polarized electrically by the action of an external electric field (i.e., the ease with which the centers of positive and negative atomic electric charge can be separated). This polarization depends on the mobility of the electrons, within the atom or molecule, in the face of resistance by molecular forces. Clearly then, ϵ_R will depend on the frequency of the applied electric field, since it will depend on how quickly these forces can respond to the field. Thus, (1.4) will be true only if n and ϵ_R refer to the same frequency of wave; hence we also note that n is frequency dependent.

1.2.2 Energy, Power, and Intensity

Let us now consider the energy content of the wave. For an electric field, the energy per unit volume, u_E, is given by (see, for example, [1])

$$u_E = \frac{1}{2}\epsilon E^2$$

and for a magnetic field

$$u_H = \frac{1}{2}\mu H^2$$

Maxwell's equations relate E and H for an electromagnetic wave according to (see Appendix A)

$$H = \left(\frac{\epsilon}{\mu}\right)^{1/2} E$$

Hence the total energy density in the wave is given by

$$u = u_E + u_H = \epsilon E^2 = \mu H^2 \tag{1.5}$$

Consider now the plane wave propagating in the direction Oz (Figure 1.1). The total energy flowing across unit area in unit time in the direction Oz will be that contained within a volume c m^3, where c is the wave velocity. Hence the power flux across unit area is given by

$$\frac{\text{power}}{\text{area}} = c\epsilon E^2 = \left(\frac{\epsilon}{\mu}\right)^{1/2} E^2$$

Clearly, if the electric field E varies sinusoidally, this quantity also will vary sinusoidally; for example, if

$$E = E_0 \cos \omega t$$

$$\frac{\text{power}}{\text{area}} = \left(\frac{\epsilon}{\mu}\right)^{1/2} \cdot E_0^2 \cos^2 \omega t = \left(\frac{\epsilon}{\mu}\right)^{1/2} \cdot \frac{1}{2} E_0^2 (1 + \cos 2\omega t)$$

The average value of this quantity over one period of oscillation is called the *intensity* of the wave (sometimes the irradiance) and clearly represents the measurable power per unit area for any device which cannot respond to optical frequencies (i.e., the vast majority). Hence we have

$$I = \left\langle \frac{\text{power}}{\text{area}} \right\rangle = \left(\frac{\epsilon}{\mu}\right)^{1/2} \langle E^2 \rangle = \left(\frac{\epsilon}{\mu}\right)^{1/2} \frac{1}{2} E_0^2 \qquad (1.6a)$$

where $\langle\ \rangle$ denotes the average value) since $\cos 2\omega t$ averages to zero.

Clearly I is proportional to the square of the electric field amplitude and also, from (1.5), it will be proportional to the square of the magnetic field amplitude. The quantity I has SI units of watts.meters^{-2}.

More generally, the intensity is expressed in terms of the Poynting vector Π (see Appendix A):

$$\Pi = \mathbf{E} \times \mathbf{H}$$

where $\mathbf{E}$ and $\mathbf{H}$ are now vector quantities and $\mathbf{E} \times \mathbf{H}$ is their vector product (see Appendix A). The intensity of the wave will be the value of Π averaged over one period of the wave. If $\mathbf{E}$ and $\mathbf{H}$ are spatially orthogonal and in phase, as in the case of a wave propagating in an isotropic dielectric medium, then

$$I = \langle \Pi \rangle = c\mu H^2 = c\epsilon E^2$$

as before. As is to be expected, in some more exotic cases (e.g., anisotropic media), the $\mathbf{E}$ and $\mathbf{H}$ components are neither orthogonal nor in phase, but $\langle \Pi \rangle$ will still provide the average power flow across unit area. If, for example, $\mathbf{E}$ and $\mathbf{H}$ happened to be in phase quadrature, then we should have

$$I = \langle \Pi \rangle = \langle E_0 \cos \omega t \cdot H_0 \sin \omega t \rangle = 0$$

and thus there is no mean power flow. (This result should be noted for reference to the case of *evanescent* waves, which will be considered later.)

In an optical medium with $\mu_R \sim 1$, (1.6a) can be written as

$$I = \left(\frac{\epsilon_R \epsilon_0}{\mu_R \mu_0}\right)^{1/2} \frac{1}{2} E_0^2 = n \left(\frac{\epsilon_0}{\mu_0}\right)^{1/2} \frac{1}{2} E_0^2 \tag{1.6b}$$

where n is, again, the refractive index of the medium. Note that I is proportional to E_0^2: this is an important relationship which will be re-emphasized shortly.

The quantity $(\mu_0/\epsilon_0)^{1/2}$ is sometimes called the *impedance of free space* and given the symbol Z_0. This is because, in free space

$$\frac{E}{H} = \left(\frac{\mu_0}{\epsilon_0}\right)^{1/2} = Z_0$$

Since E has dimensions of volts.meters^{-1} and H of amps.meters^{-1}, Z_0 clearly has the dimensions of impedance (ohms); Z_0 is real and has the value

$$\left(\frac{\mu_0}{\epsilon_0}\right)^{1/2} = \left(\frac{4\pi \times 10^{-7}}{8.854 \times 10^{-12}}\right)^{1/2} = 376.7 \text{ ohms}$$

It follows that (1.6b) can be written as

$$I = \frac{n}{2Z_0} E_0^2 = \frac{n}{753.46} E_0^2 = 1.33 \times 10^{-3} n E_0^2 \tag{1.6c}$$

This is a useful relationship in two ways. First, it relates a quantity that is directly measurable (I) with one which is not (E_0). Second, it provides the actual numerical relationship between I and E_0, and this is valuable when designing devices and systems, as we shall discover later.

1.2.3 Optical Polarization

We should now give brief consideration to what is known as the *polarization* of the optical wave. (This topic will be dealt with much more comprehensively in Chapter 3.)

The "typical" sinusoidal solution of Maxwell's wave equation given by (1.1) is, of course, only one of an infinite number of such sinusoidal solutions. The general solution for a sinusoid of angular frequency ω is given by

$$\mathbf{E}(\mathbf{r}, t) = \mathbf{E}(\mathbf{r}) \exp(i\omega t)$$

where $\mathbf{E}(\mathbf{r}, t)$, $\mathbf{E}(\mathbf{r})$ are, in general, complex vectors, and $\mathbf{r}$ is a real radius vector in the xy plane.

If, for simplicity, we consider just plane, monochromatic (single frequency) waves propagating in free space in the direction Oz, we may, for the $\mathbf{E}$ field, write the general solution to the wave equation in the form

$$E_x = e_x \cos(\omega t - kz + \delta_x)$$

$$E_y = e_y \cos(\omega t - kz + \delta_y)$$

where δ_x, δ_y, are arbitrary (but constant) phase angles. Thus, we are able to describe this solution completely by means of two waves: one in which the electric field lies entirely in the xz plane, and the other in which it lies entirely in the yz plane (Figure 1.2). If these waves are observed at a particular value of z, say, z', then they take the oscillatory form

$$E_x = e_x \cos(\omega t + \delta'_x); \quad \delta'_x = \delta_x - kz'$$

$$E_y = e_y \cos(\omega t + \delta'_y); \quad \delta'_y = \delta_y - kz'$$

and the tip of each vector appears to oscillate sinusoidally with time along a line. E_x is said to be linearly polarized in the direction Ox, and E_y linearly polarized in the direction Oy.

The tip of the vector which is the sum of E_x and E_y will, in general, describe an ellipse whose Cartesian equation in the xy plane at the chosen z' will be given by eliminating ωt from the expression for E_x and E_y; that is,

$$\frac{E_x^2}{e_x^2} + \frac{E_y^2}{e_y^2} - \frac{2E_xE_y}{e_xe_y}\cos\delta = \sin^2\delta$$

$$\delta = \delta'_y - \delta'_x$$

This ellipse will degenerate into a straight line (and the overall polarization state of the light will thus be linear) if

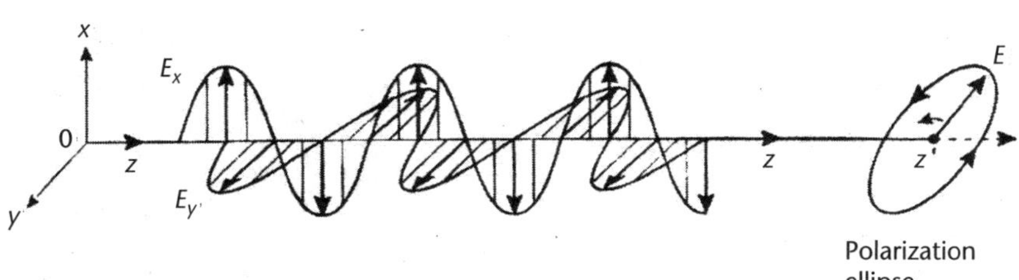

Figure 1.2 Electric field components for an elliptically polarized wave.

$$e_x \neq 0; \; e_y = 0$$

or

$$e_x = 0; \; e_y \neq 0$$

or

$$\delta = m\pi$$

where m is any integer, including zero. This corresponds to the condition that E_x and E_y are either in phase or in antiphase.

The ellipse becomes a circle (and the light is thus circularly polarized) if

$$e_x = e_y$$

and

$$\delta = (2m + 1)\,\pi/2$$

that is, the waves are equal in amplitude and are in phase quadrature.

The polarization properties of light waves are especially important for propagation within anisotropic media, in which the physical properties vary with direction. In this case the propagation characteristics for the component E_x will, in general, differ from those for E_y, so that the values of e_x, e_y and δ will vary along the propagation path. The polarization state of the light will now become dependent upon the propagation distance, and on the state of the medium. This, also, will be covered in detail in Chapter 3.

1.3 Reflection and Refraction

We have seen in Section 1.2 that Maxwell's equations allow a set of solutions of the form

$$E_x = E_0 \exp[i(\omega t - kz)]$$
$$H_y = H_0 \exp[i(\omega t - kz)]$$

with $\omega/k = (\epsilon\mu)^{-1/2} = c$.

These represent plane waves traveling in the Oz direction. We shall now investigate the behavior of such waves, with particular regard to the effects that occur at the boundaries between different optical media.

Of course, other types of solution are also possible. An important solution is that of a wave which spreads spherically from a point to a distance r:

$$E_r = \frac{E_0}{r} \exp\left[i(\omega t - kr)\right]$$

Here the factor $1/r$ in the amplitude is necessary to ensure conservation of energy (via the Poynting vector) for, clearly, the total area over which the energy flux occurs is $4\pi r^2$, so that the intensity falls as $1/r^2$. (Remember that intensity is proportional to the square of the amplitude.)

It is interesting and valuable to note that the propagation of a plane wave (such as in Figure 1.3) is equivalent to the propagation of spherical waves radiating from each point on the propagating wavefront of the plane wave. On a given wavefront the waves at each point begin in phase (this is the definition of a wavefront), so that they remain strictly in phase only in a direction at right angles to the front (Figure 1.3). Hence the plane wave appears to propagate in that direction. This principle of equivalence, first enunciated by Huygens and later shown by Kirchhoff to be mathematically sound [2], is very useful in the study of wave propagation phenomena generally.

The laws of reflection and refraction were first formulated in terms of *rays* of light. It had been noticed (around 1600) that, when dealing with *point* sources, the light passed through apertures consistently with the view that it was composed of rays traveling in straight lines from the point. (It was primarily this observation that led to Newton's *corpuscular* theory.) The practical concept was legitimized by allowing such light to pass through a small hole so as to

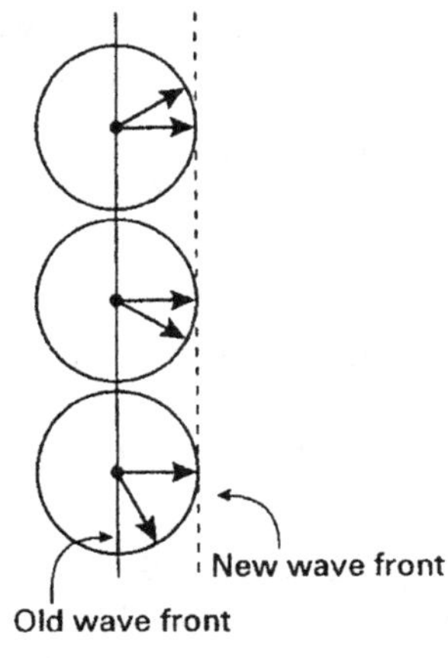

Figure 1.3 Huygens' construction.

isolate a *ray*. Such rays were produced, and their behavior in respect of reflection and refraction at material boundaries was formulated, thus:

1. On reflection at a boundary between two media, the reflected ray lies in the same plane as that of the incident ray and the normal to the boundary at the point of incidence (the plane of incidence); the angle of reflection equals the angle of incidence.
2. On refraction at a boundary, the refracted ray also lies in the plane of incidence, and the sine of the angle of refraction bears a constant ratio to the sine of the angle of incidence (Snell's law).

These two laws form the basis of what is known as geometrical optics, or, ray optics. The majority of bulk optics (e.g., lens design, reflectometers, prismatics) can be formulated with its aid. However, it has severe limitations. For example, it cannot predict the intensities of the refracted and reflected rays.

If, in the attempt to isolate a ray of light of increasing fineness, the aperture is made too small, the ray divergence appears to increase, rather than diminish. This occurs when the aperture size becomes comparable with the wavelength of the light, and it is under this condition that the geometrical theory breaks down. *Diffraction* has occurred and this is, quintessentially, a wave phenomenon. The wave theory provides a more complete, but necessarily more complex, view of light propagation. We shall now deal with the phenomena of reflection and refraction using the wave theory, but we should remember that, under certain conditions (apertures much larger than the wavelength), the ray theory is useful for its simplicity: a wave can be replaced by a set of rays in the direction of propagation, normal to surfaces of constant phase, and obeying simple geometrical rules.

Let us consider two nonconducting dielectric media with refractive indices n_1 and n_2, separated by a plane boundary which we take to be the xy plane at $z = 0$ (Figure 1.4). Let us now consider a plane wave lying in the xz plane, which is propagating in medium 1 and is incident on the boundary at angle ϑ_i, as shown in the figure. All the field components, such as (E_i, H_i), will vary as

$$(E_i, H_i) \exp\{i\omega[t - n_1(x \sin \vartheta_i + z \cos \vartheta_i)/c]\}$$

(see Figure 1.5) using the exponential forms of the wave and taking c to be the velocity of light in free space.

After striking the boundary there will, in general, be a reflected and a refracted (transmitted, t) wave. This fact is a direct consequence of the boundary conditions that must be satisfied at the interface between the two media. These conditions follow from Maxwell's equations, and essentially may be stated as

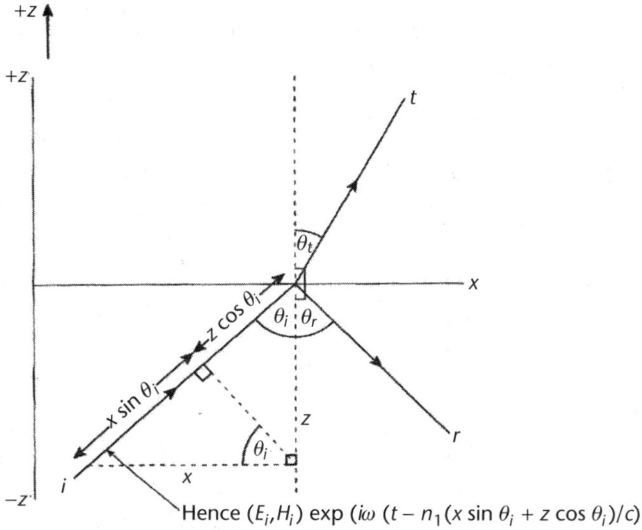

Figure 1.4 Reflection and refraction at a boundary between two media.

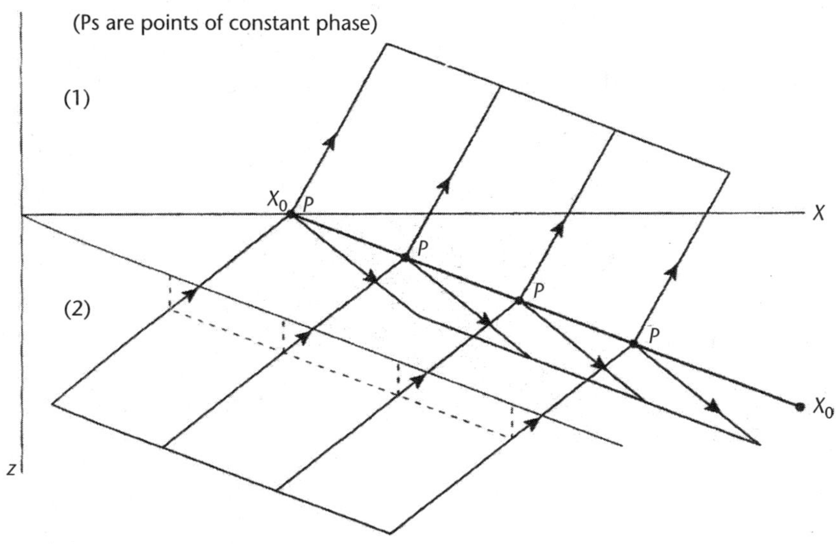

Figure 1.5 Line of constant phase in boundary plane.

1. Tangential components of **E** and **H** are continuous across the boundary.
2. Normal components of **B** and **D** are continuous across the boundary.

The above conditions must be true at all times and at all places on the boundary plane. They can only be true at all times at a given point if the frequencies of all the waves (i.e., incident, reflected, refracted) are the same; otherwise, clearly, amplitude discontinuities would occur across the boundary. Furthermore, since the phase and amplitude of the incident wave must be constant on the boundary plane along any line for which x is constant (see Figure 1.5), it follows that the phases and amplitudes of the reflected and refracted waves must also be constant along such a line, if continuity in accordance with the boundary conditions is to be maintained, and this is equivalent to saying that the reflected and refracted rays travel in the same direction and thus in the same plane (the xz plane) as the incident ray, which proves one of the previously stated laws of reflection and refraction.

To go further it is necessary to give proper mathematical expression to the waves. Any given wave is, of course, a sinusoid, whose amplitude, frequency and phase define the wave completely, and the most convenient representation of such waves is via their complex exponential form.

Suppose (Figure 1.6) that the reflected and refracted waves make angles ϑ_r and ϑ_t, respectively, to the boundary in the xz plane. Then these waves will vary as

$$\text{reflected: } \exp\{i\omega[t - n_1(x \sin \vartheta_r - z \cos \vartheta_r)/c]\}$$

(note that the reflected ray travels in the *negative* z direction)

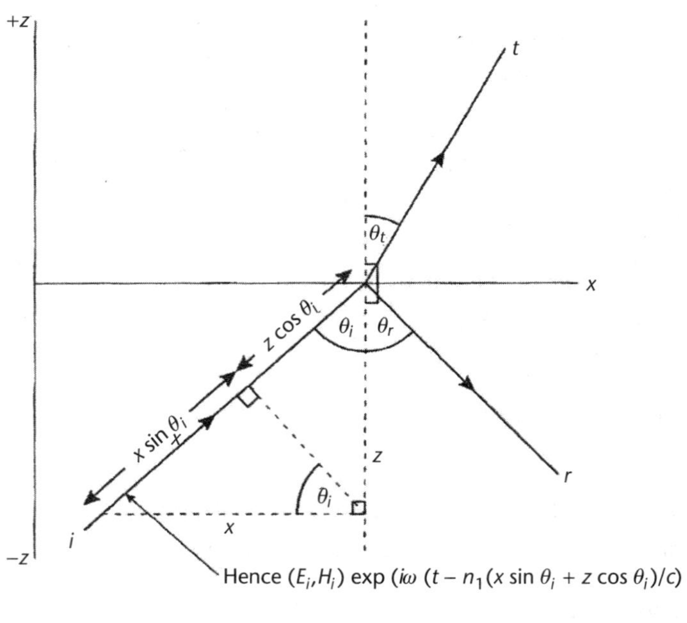

Figure 1.6 Trigonometry of the incident ray.

$$\text{refracted: } \exp\{i\omega[t - n_2(x \sin \vartheta_t + z \cos \vartheta_t)/c]\}$$

whereas the incident wave, for reference, was

$$\text{incident: } \exp\{i\omega[t - n_1(x \sin \vartheta_i + z \cos \vartheta_i)/c]\}$$

At the boundary ($z = 0$), these variations must be identical for any x, t, if continuity is to be maintained, hence,

$$n_1 x \sin \vartheta_i = n_1 x \sin \vartheta_r = n_2 x \sin \vartheta_t$$

Thus, we have

$$\vartheta_i = \vartheta_r \text{ (law of relection)}$$

$$n_1 \sin \vartheta_i = n_2 \sin \vartheta_t \text{ (Snell's law of refraction)}$$

We must now consider the relative amplitudes of the waves. To do this we match the components of **E**, **H**, **D**, **B**, separately. A further complication is that the values of these quantities at the boundary will depend on the direction of vibration of the **E**, **H**, fields of the incident wave, relative to the plane of the wave. Therefore, we need to consider two linear, orthogonal polarization components separately, one in the *xz* plane, the other normal to it. (Any other polarization state can be resolved into these two linear components, so that our solution will be complete.)

Let us consider the two stated linear components in turn.

E *in the plane of incidence;* **H** *normal to the plane of incidence*

The incident wave can now be written in the form (see Figure 1.6)

$$\begin{aligned} E_x^i &= -E_i \cos \vartheta_i \exp\{i\omega[t - n_1(x \sin \vartheta_i + z \cos \vartheta_i)/c]\} \\ E_z^i &= E_i \sin \vartheta_i \exp\{i\omega[t - n_1(x \sin \vartheta_i + z \cos \vartheta_i)/c]\} \qquad (1.7) \\ H_y^i &= H_i \exp\{i\omega[t - n_1(x \sin \vartheta_i + z \cos \vartheta_i)/c]\} \end{aligned}$$

Now we can again enlist the help of Maxwell's equations to relate **H** and **E** for a plane wave (see Appendix A).

We have

$$\frac{E}{H} = Z = \left(\frac{\mu}{\epsilon}\right)^{1/2}$$

Z is now known as the characteristic impedance of the medium. Since we are dealing, in this case, with nonconducting dielectrics, we have $\mu = 1$ and $n = \epsilon^{1/2}$; hence,

$$Z = \frac{Z_0}{n}$$

Thus,

$$H_i = nE_i/Z_0 \tag{1.8}$$

and the expression for H_y^i becomes

$$H_y^i = n_1 E_i/Z_0 \cdot \exp\{i\omega[t - n_1(x \sin \vartheta_i + z \cos \vartheta_i)/c]\}$$

Clearly we can construct similar sets of equations for the reflected and refracted waves. Having done this we can impose the boundary conditions to obtain the required relationships between wave amplitudes. We shall now derive these relationships—that is, that between the reflected and incident electric field amplitudes, and that between the refracted and incident electric field amplitudes for this case.

We know that the exponential factors are all identical at the boundary if we are going to be able to satisfy the boundary conditions at all; let us, therefore, write the universal exponential factor as F.

For the incident (i) wave, from (1.7) we have

$$\begin{aligned} E_x^i &= -E_i \cos \vartheta_i F \\ E_z^i &= E_i \sin \vartheta_i F \\ H_y^i &= H_i F \end{aligned} \tag{i}$$

For the reflected (r) wave:

$$\begin{aligned} E_x^r &= E_r \cos \vartheta_r F \\ E_z^r &= E_r \sin \vartheta_r F \\ H_y^r &= H_r F \end{aligned} \tag{r}$$

For the refracted (t) wave:

$$E_x^t = -E_r \cos \vartheta_r F \qquad (t)$$
$$E_z^t = E_t \sin \vartheta_t F$$
$$H_y^t = H_t F$$

Imposing the condition that the tangential components (i.e., x components) of E must be continuous across the boundary, we have

$$E_x^i + E_x^r = E_x^t$$

or

$$-E_i \cos \vartheta_i + E_r \cos \vartheta_r = E_t \cos \vartheta_t \qquad (1.9)$$

using the appropriate equations from (i), (r) and (t) and canceling the factor F.

Now doing the same for the tangential H field (y components):

$$H_i + H_r = H_t \qquad (1.10)$$

We also know, from (1.8), that

$$H_i = n_1 E_i / Z_0; \; H_r = n_1 E_r / Z_0; \; H_t = n_2 E_t / Z_0$$

hence the H field condition (1.10) becomes

$$n_1 E_i + n_1 E_r = n_2 E_t \qquad (1.11)$$

We may now eliminate E_t from (1.9) and (1.11) to obtain (remembering, also, that $\vartheta_r = \vartheta_i$)

$$\frac{E_r}{E_i} = \frac{n_2 \cos \vartheta_i - n_1 \cos \vartheta_t}{n_2 \cos \vartheta_i + n_1 \cos \vartheta_t} \qquad (1.12a)$$

which is the required relationship.

Note also that, since, from Snell's law, $n_1 \sin \theta_i = n_2 \sin \theta_t$, this can be written as

$$\frac{E_r}{E_i} = \frac{\tan(\vartheta_i - \vartheta_t)}{\tan(\vartheta_i + \vartheta_t)}$$

Similarly we may eliminate E_r from (1.9) and (1.11) to obtain

$$\frac{E_t}{E_i} = \frac{2n_1 \cos \vartheta_t}{n_2 \cos \vartheta_i + n_1 \cos \vartheta_t} \tag{1.12b}$$

We must now consider the wave with the other, orthogonal, polarization.

E *normal to the plane of incidence;* **H** *in the plane of incidence*

Using the same methods as before we obtain the relations

$$\frac{E'_r}{E'_i} = \frac{n_1 \cos \vartheta_i - n_2 \cos \vartheta_t}{n_1 \cos \vartheta_i + n_2 \cos \vartheta_t} \tag{1.12c}$$

$$\frac{E'_t}{E'_i} = \frac{2n_1 \cos \vartheta_i}{n_1 \cos \vartheta_i + n_2 \cos \vartheta_t} \tag{1.12d}$$

The above four expressions (1.12) are known as *Fresnel's equations*; Fresnel derived them from the elastic-solid theory of light, which prevailed at his time. The equations contain several points worthy of emphasis.

First, we note that there is a possibility of eliminating the reflected wave. For E in the plane of incidence, we find from (1.12a) that this occurs when

$$n_1 \cos \vartheta_t = n_2 \cos \vartheta_i$$

But from Snell's law we also have

$$n_1 \sin \vartheta_i = n_2 \sin \vartheta_t$$

so that, combining the two relations

$$\sin 2\vartheta_i = \sin 2\vartheta_t$$

Now, of course, this equation has an infinite number of solutions, but the only one of interest is that for which $\vartheta_i \neq \vartheta_t$ ($\vartheta_i = \vartheta_t$ only if $n_1 = n_2$) and for which both ϑ_i and ϑ_t lie in the range $0 \to \pi/2$. The required solution is

$$\vartheta_i + \vartheta_t = \frac{1}{2}\pi$$

and simple geometry then requires that the reflected and refracted rays are normal to each other (Figure 1.7). Clearly, from Snell's law, this occurs when

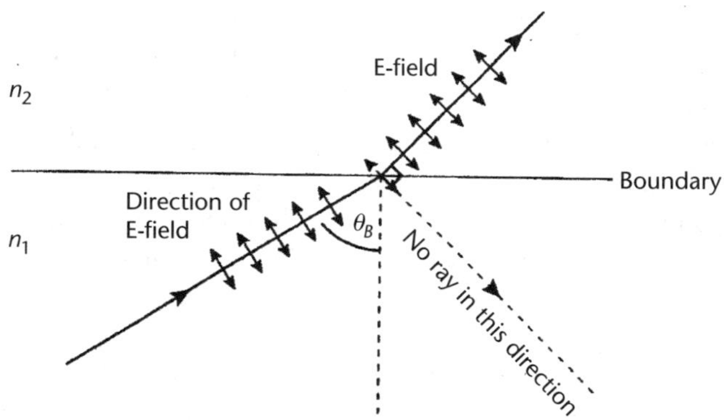

Figure 1.7 Elimination of the reflected ray at the Brewster angle (ϑ_B).

$$n_1 \sin \vartheta_i = n_2 \cos \vartheta_i$$

that is,

$$\tan \vartheta_i = \frac{n_2}{n_1}$$

This particular value of ϑ_i is known as Brewster's angle (ϑ_B).

For example, for the glass/air boundary we find $\vartheta_B = 56.3°$.

It is instructive to understand the physical reason for the disappearance of the reflected ray at this angle when the electric field lies in the plane of incidence. Referring to Figure 1.7 we note that the incident wave sets up oscillations of the elementary dipoles in the second medium and, at the Brewster angle, these oscillations take place in the direction of the reflected ray, since the refracted and reflected rays are orthogonal. Hence these oscillations cannot generate any transverse waves in the required direction of reflection. Since light waves are, by their very nature, transverse, the reflected ray must be absent. If we ask the same question of the polarization that has **E** normal to the plane of incidence, we find from (1.12c) that

$$n_1 \cos \vartheta_i = n_2 \cos \vartheta_t$$

which, with Snell's law, gives

$$\tan \vartheta_i = \tan \vartheta_t$$

There is no solution of this equation that satisfies the required conditions, so the reflected wave cannot be eliminated in this case. If, then, a wave of arbitrary polarization is incident on the boundary at the Brewster angle, only the polarization with **E** normal to the plane of incidence is reflected. This is a useful way of linearly polarizing a wave.

The second point worthy of emphasis is the condition at normal incidence. Here we have $\vartheta_i = \vartheta_r = \vartheta_t = 0$; hence the relations, identical for both polarizations, become

$$\frac{E_r}{E_i} = \frac{E'_r}{E'_i} = \frac{n_1 - n_2}{n_1 + n_2} \tag{1.13a}$$

$$\frac{E_t}{E_i} = \frac{E'_t}{E'_i} = \frac{2n_1}{n_1 + n_2} \tag{1.13b}$$

Now the wave intensities are proportional to the squares of the electric field amplitudes *but only for a given medium*, since, from (1.6c), the intensity is proportional to the refractive index as well as to the square of the field.

Hence, since the incident and reflected waves propagate in the same medium, it is appropriate to write

$$\frac{I_r}{I_t} = \frac{E_r^2}{E_t^2} = \left(\frac{n_1 - n_2}{n_1 + n_2}\right)^2 \tag{1.13c}$$

but for the transmitted (refracted) wave, we have

$$\frac{I_t}{I_i} = \frac{n_2 E_t^2}{n_1 E_i^2} = \frac{4n_1 n_2}{(n_1 + n_2)^2} \tag{1.13d}$$

Note that now:

$$I_r + I_t = I_i$$

so that energy is conserved, as required.

Equations (1.13c) and (1.13d) are useful expressions, for they tell us how much light is lost by normal reflection when transmitting from one medium (say, air) to another (say, glass). For example, when passing through a glass lens (air → glass → air), taking the refractive index of the glass as 1.5 we find from (1.13c) that the fractional loss at the front face of the lens (assumed approximately normal) is

$$\frac{I_r}{I_i} = \frac{(0.5)^2}{(2.5)^2} = 0.04$$

Another 4% will be lost at the back face, giving a total Fresnel loss of the order of 8%. This figure can be reduced by *antireflection* coatings.

Finally we should notice that all the expressions for the ratios of field amplitudes are mathematically real, and thus any change of phase which occurs at a boundary must be either 0 or π. We shall now look at a rather different type of reflection where this is not the case.

1.4 Total Internal Reflection

We return to Snell's law:

$$n_1 \sin \vartheta_i = n_2 \sin \vartheta_t$$

or

$$\sin \vartheta_t = \frac{n_1}{n_2} \sin \vartheta_i \quad (1.14)$$

The factor $\sin \vartheta_i$ is, of course, always less than unity. However if $n_2 < n_1$ (i.e., the second medium is less optically dense than the first, which contains the incident ray) then it may be that

$$\sin \vartheta_i > \frac{n_2}{n_1}$$

that is,

$$\frac{n_1}{n_2} \sin \vartheta_i > 1$$

If this is so, then we have from (1.14)

$$\sin \vartheta_t > 1 \quad (1.15)$$

Equation (1.15) clearly cannot be satisfied for any real value of ϑ_t and there can be no real refracted ray. The explanation of this is that the refracted ray angle (ϑ_t), under these conditions of passage from a less dense to a more dense medium, is always greater than the incident angle (ϑ_i). Consequently ϑ_t will reach a value of 90° (i.e., parallel to the boundary) before ϑ_i, and any greater value of ϑ_i cannot yield a refracted ray (Figure 1.8). The value of ϑ_i for which (1.15) just becomes true, we define as the critical angle, ϑ_c

$$\sin \vartheta_c = \frac{n_2}{n_1}$$

For all values of $\vartheta_i > \vartheta_c$, the light is totally reflected at the boundary: the phenomenon is called total internal reflection (TIR). However, Fresnel's equations must still apply, for we made no limitations on the values of the quantities when imposing the boundary conditions. Furthermore, if the fields are to be continuous across the boundary, as required by Maxwell's equations, then there must be a field disturbance of some kind in the second medium. We can use Fresnel's equations to investigate this disturbance.

We write:

$$\cos \vartheta_t = (1 - \sin^2 \vartheta_t)^{1/2} \tag{1.16}$$

Since $\sin \vartheta_t > 1$ for $\vartheta_t > \vartheta_c$, and since also the function $\cosh \gamma \geq 1$ for all real γ, we may, for convenience, use the substitution

$$\sin \vartheta_t = \cosh \gamma \; (\vartheta_i > \vartheta_c)$$

and henceforth, therefore, the TIR condition (1.15) is, implicitly, imposed. We now have, from (1.16),

$$\cos \vartheta_t = i(\cosh^2 \gamma - 1)^{1/2} = \pm i \sinh \gamma$$

Hence we may write the field components in the second medium to vary as

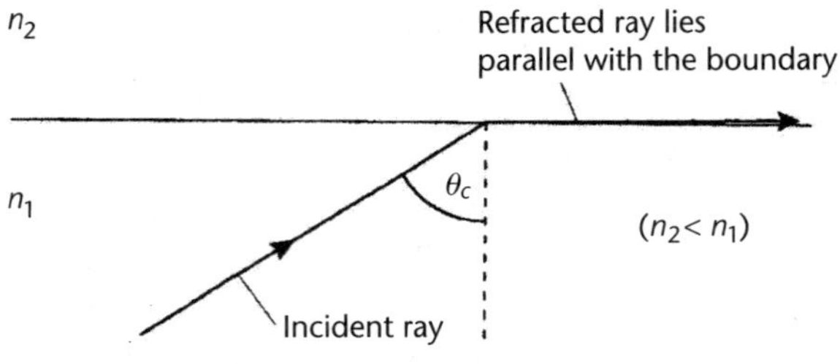

Figure 1.8 Critical angle (ϑ_c) for total internal reflection (TIR).

$$\exp\{i\omega[t - n_2(x \cosh\ \gamma - iz \sinh\ \gamma)/c]\}$$

or

$$\exp[(-\omega n_2 z \sinh\ \gamma)/c]\ \exp[i\omega(t - n_2 x \cosh\ \gamma)/c]$$

where we have used the fact that $\cosh\ \gamma = \frac{1}{2}(e^{\gamma} + e^{-\gamma})$ which tends to infinity as $\gamma \rightarrow +\infty$, and has a minimum of 1 at $\gamma = 0$.

This represents a wave traveling in the *Ox* direction in the second medium (i.e., parallel to the boundary) with amplitude decreasing exponentially in the *Oz* direction (at right angles to the boundary). The rate at which the amplitude falls with z can be written:

$$\exp[(-2\pi z \sinh\ \gamma)/\lambda_2]$$

or, in terms of the original parameters:

$$\exp\left[-k_2 z\left(n_1^2 \sin^2\ \vartheta_i - n_2^2\right)^{1/2}/n_2\right]$$

λ_2 being the wavelength of the light and k_2 the wave-number, in the second medium. This shows that the wave is attenuated significantly over distances $\sim \lambda_2$. For example, at the glass/air interface, the critical angle will be $\sim \sin^{-1}$ (1/1.5) (i.e., ~ 42°). For a wave in the glass incident on the glass/air boundary at 60° ($\vartheta_i > \vartheta_c$), we find that $\sinh\ \gamma = 1.64$. Hence the amplitude of the wave in the second medium is reduced by a factor of 5.4×10^{-3} in a distance of only one wavelength, the latter being of order 1 μm. The wave is called an *evanescent* wave. Even though the evanescent wave is propagating in the second medium, it transports no light energy in a direction normal to the boundary. All the light is totally internally reflected at the boundary. The fields which exist in the second medium give a Poynting vector which averages to zero in this direction, over one oscillation period of the light wave. All the energy in the evanescent wave is transported parallel to the boundary between the two media. The totally internally reflected wave now suffers a phase change which depends both on the angle of incidence and on the polarization. This can readily be derived from Fresnel's equations. Taking (1.12a) we have for the TIR case where **E** lies in the plane of incidence:

$$\frac{E_r}{E_i} = \frac{n_2 \cos\ \vartheta_i - in_1 \sinh\ \gamma}{n_2 \cos\ \vartheta_i + in_1 \sinh\ \gamma}$$

This complex number provides the phase change on TIR as δ_p where

$$(E_{para})\colon \tan\left(\frac{1}{2}\delta_p\right) = \frac{n_1\left(n_1^2 \sin^2 \vartheta_i - n_2^2\right)^{1/2}}{n_2^2 \cos \vartheta_i}$$

and for the perpendicular E polarization:

$$(E_{perp})\colon \tan\left(\frac{1}{2}\delta_s\right) = \frac{\left(n_1^2 \sin^2 \vartheta_i - n_2^2\right)^{1/2}}{n_1 \cos \vartheta_i}$$

We note also that

$$\tan\left(\frac{1}{2}\delta_p\right) = n_1^2 \tan\left(\frac{1}{2}\delta_s\right)$$

and that

$$\tan\left[\frac{1}{2}(\delta_p - \delta_s)\right] = \frac{\cos \vartheta_i\left(n_1^2 \sin^2 \vartheta_i - n_2^2\right)^{1/2}}{n_1 \sin^2 \vartheta_i}$$

The variations δ_p, δ_s and $\delta_p - \delta_s$ are shown in Figure 1.9 as a function of ϑ_i. It is clear that the polarization state of light undergoing TIR will be changed as a result of the differential phase change $\delta_p - \delta_s$. By choosing ϑ_i appropriately and, perhaps, using two TIRs, it is possible to produce any wanted, final polarization state from any given initial state.

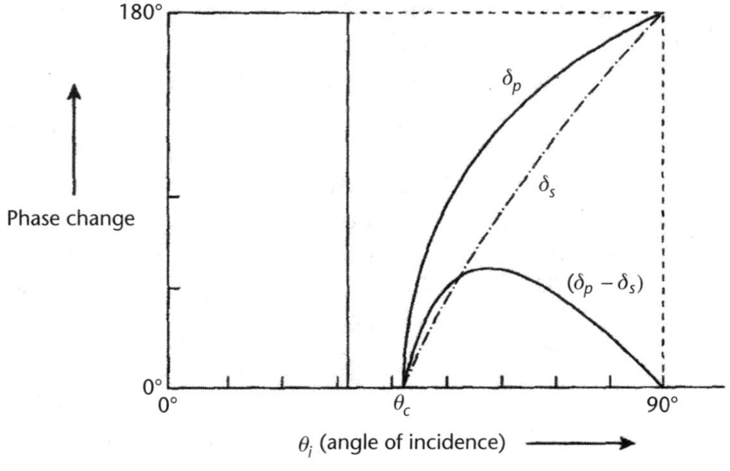

Figure 1.9 Phase changes on total internal reflection.

It is interesting to note that the reflected ray in TIR appears to originate from a point which is displaced along the boundary from the point of incidence. This is consistent with the incident ray being reflected from a parallel plane that lies a short distance within the second boundary (Figure 1.10). This view is also consistent with the observed phase shift, which now is regarded as being due to the extra optical path traveled by the ray. The displacement is known as the Goos-Hanchen effect and provides an entirely consistent alternative explanation of TIR. This provides food for further interesting thoughts, which we shall not pursue since they are somewhat beyond the scope of this book.

1.5 Interference of Light

We have seen that light consists of oscillating electric and magnetic fields. We know that these fields are vector fields since they represent forces (on unit charge and unit magnetic pole, respectively). The fields will thus add vectorially. Consequently, when two light waves are superimposed on each other, we obtain the resultant by constructing their vector sum at each point in time and space, and this fact has already been used in consideration of the polarization of light (Section 1.2.3).

If two sinusoids are added, the result is another sinusoid. Suppose that two light waves given, via their electric fields, as

$$e_1 = E_1 \cos(\omega t + \varphi_1)$$
$$e_2 = E_2 \cos(\omega t + \varphi_2)$$

have the same polarization and are superimposed at a point in space. We know that the resultant field at the point will be given, using elementary trigonometry, by

$$e_T = E_T \cos(\omega t + \varphi_T)$$

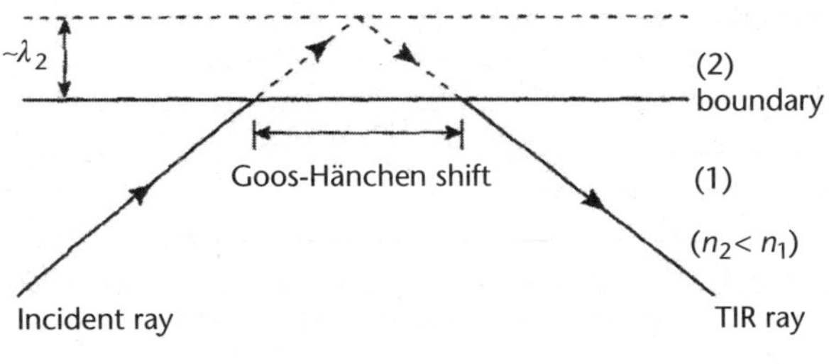

Figure 1.10 The Goos-Hanchen shift on total internal reflection.

where

$$E_T^2 = E_1^2 + E_2^2 + 2E_1E_2 \cos(\varphi_2 - \varphi_1)$$

and

$$\tan \vartheta_T = \frac{E_1 \sin \varphi_1 + E_2 \sin \varphi_2}{E_1 \cos \varphi_1 + E_2 \cos \varphi_2}$$

For the important case where $E_1 = E_2 = E$, say, we have

$$E_T^2 = 4E^2 \cos^2 \frac{1}{2}(\varphi_2 - \varphi_1) \tag{1.17}$$

and

$$\tan \phi_T = \tan \frac{1}{2}(\varphi_2 + \varphi_1)$$

The intensity of the wave will be proportional to E_T^2 so that, from (1.17) it can be seen to vary from $4E^2$ to 0, as $(\varphi_2 - \varphi_1)/2$ varies from 0 to $\pi/2$.

Consider now the arrangement shown in Figure 1.11. Here two slits, separated by a distance p, are illuminated by a plane wave with wavelength λ. The portions of the wave that pass through the slits will interfere on the screen S, a distance d away. Now each of the slits will act as a source of cylindrical

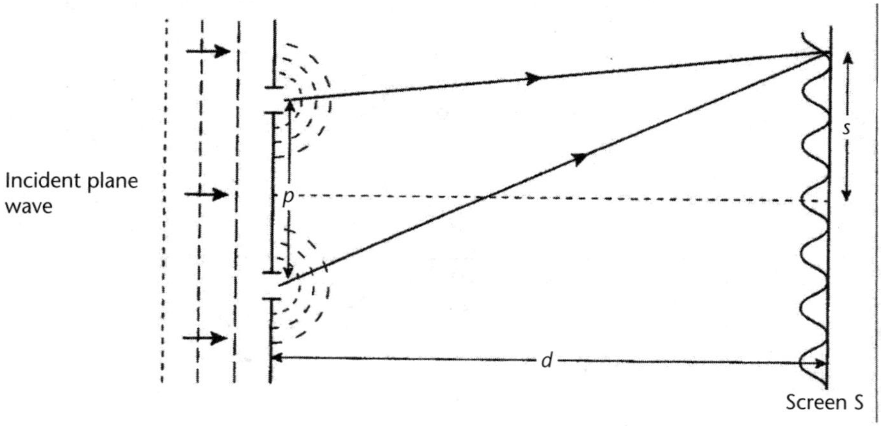

Figure 1.11 "Young's slits" interference.

waves, from Huygens' principle. Moreover, since they originate from the same plane wave, they will start in phase. On a line displaced a distance s from the line of symmetry on the screen, the waves from the two slits will differ in phase by

$$\delta = \frac{2\pi}{\lambda}\frac{sp}{d} \qquad (d \gg s, p)$$

Thus, as s increases, the intensity will vary between a maximum and zero, in accordance with (1.17). These variations will be viewed as fringes (i.e., lines of constant intensity parallel with the slits). They are known as Young's fringes, after their discoverer, and are the simplest example of light interference. Such interference is an essential feature of any wave motion, and light interference effects and phenomena pervade the whole of optical physics, phenomena which can be used in a wide variety of complex and, sometimes, quite subtle ways.

1.6 Diffraction

In Section 1.3 it was noted that each point on a wavefront could be regarded formally and rigorously as a source of spherical waves. In Section 1.5 it was noted that any two waves, when superimposed, will interfere. Consequently wavefronts can interfere with themselves and with other, separate, wavefronts. To the former usually is attached the name *diffraction*, and to the latter *interference*, but the distinction is somewhat arbitrary and, in several cases, far from clear-cut.

Diffraction of light may be regarded as the limiting case of multiple interference as the source spacings become infinitesimally small. Consider the slit aperture in Figure 1.12. This slit is illuminated with a uniform plane wave and the light which passes through the slit is observed on a screen which is sufficiently distant from the slit for the light which falls upon it to be effectively, again, a plane wave. These are the conditions for Fraunhofer diffraction. If source and screen are close enough to the slit for the waves not to be plane, we have a more complex situation, known as Fresnel diffraction. Fraunhofer diffraction is by far the more important of the two, and is the only form of diffraction we shall deal with here. Fresnel diffraction usually can be transformed into Fraunhofer diffraction, in any case, by the use of lenses which render the waves effectively plane, even over short distances.

Suppose that in Figure 1.12 the amplitude of the wave at distances between x and $x + dx$ along the slit is given by the complex quantity $f(x)\,dx$, and consider the effect of this at angle ϑ, as shown. (Since each point on the wavefront acts as a source of spherical waves, all angles will, of course, be

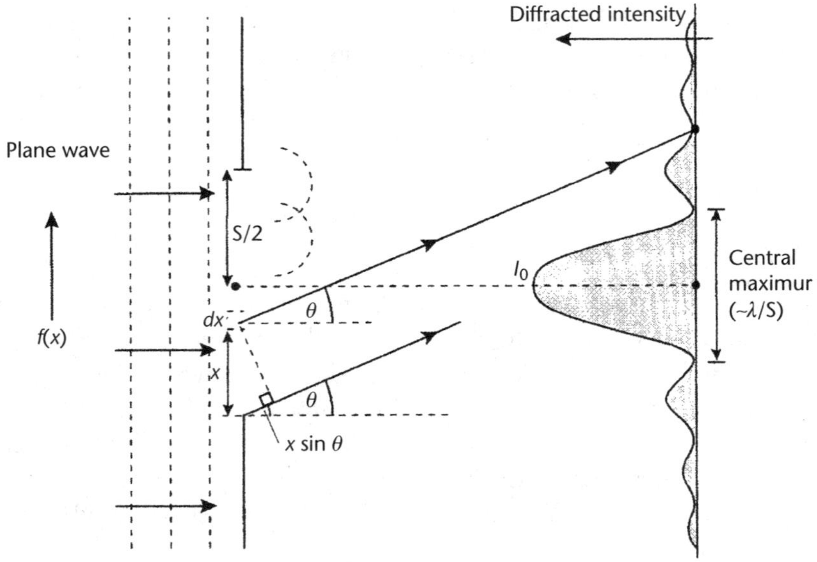

Figure 1.12 Diffraction of a slit.

illuminated by the strip.) The screen, being effectively infinitely distant from the slit, will be illuminated at one point by the light leaving the slit at angles between ϑ and $\vartheta + d\vartheta$. Taking the bottom of the slit as the phase reference, the light, on arriving at the screen, will lead by a phase

$$\Phi = kx \sin \vartheta$$

and hence the total amplitude in directions ϑ to $\vartheta + d\vartheta$ will be given by

$$A(\vartheta) = \int_{-\infty}^{\infty} f(x) \exp(-ikx \sin \vartheta)\, dx$$

We can also write

$$A(\alpha) = \int_{-\infty}^{\infty} f(x) \exp(-i\alpha x)\, dx$$

with

$$\alpha = k \sin \vartheta$$

Hence $A(\alpha)$ and $f(x)$ constitute a reciprocal Fourier transform pair (see Appendix B); that is, each is the Fourier transform of the other. This is an important result. For small values of ϑ it implies that the angular distribution of the diffracted light is the Fourier transform of the aperture's amplitude distribution.

Let us see how this works for some simple cases. Take first a uniformly illuminated slit of width s. The angular distribution of the diffracted light will now be

$$A(k \sin \vartheta) = \int_{-\pi/2}^{\pi/2} a \exp(-ikx \sin \vartheta)\, dx$$

where a is the (uniform) amplitude at the slit per unit of slit width.

Hence,

$$A(k \sin \vartheta) = a \frac{\sin\left(\frac{1}{2} ks \sin \vartheta\right)}{\left(\frac{1}{2} ks \sin \vartheta\right)}$$

Writing, for convenience, $\beta = \frac{1}{2} ks \sin \vartheta$, we find that the intensity in a direction ϑ is given by

$$I(\vartheta) = (as)^2 \frac{\sin^2 \beta}{\beta^2} = I_0 \frac{\sin^2 \beta}{\beta^2} \tag{1.18}$$

where I_0 is intensity at the center of the diffraction pattern. This variation is shown in Figure 1.12 and, as in the case of multiple interference between discrete sources, its shape is a result of the addition of wave vectors with phase increasing steadily with ϑ.

This form of variation occurs frequently in physics across a broad range of applications and it is instructive to understand why. The function appropriate to the variation is given the name "sinc" (pronounced "sink"); that is,

$$\text{sinc}(\beta) = \frac{\sin \beta}{\beta}$$

$$\text{sinc}^2(\beta) = \frac{\sin^2 \beta}{\beta^2}$$

Let us examine the physical reason for the sinc function in the case we have been considering (i.e., a uniformly illuminated slit). In this case each infinitesimal element of the slit provides a wave amplitude adx and at the center of the screen all of these elements are in phase, producing a total amplitude, *as*. Hence it is possible to represent all these elementary vectors as a straight line (since they are all in phase) of length *as* [Figure 1.13(a)]. Now consider the situation at angle ϑ to the axis. As already shown the ray from the bottom of the slit lags that from the top by a phase

$$\varphi_T = ks \sin \vartheta = 2\beta$$

The result can, therefore, be depicted as in Figure 1.13(b). The first and last infinitesimal vectors are inclined at 2β to each other and the intervening vectors form an arc of a circle that subtends 2β at the circle's center. The vector addition of all the vectors thus leads to a resultant that is the chord across the arc in Figure 1.13(b). Simple geometry gives the length of this chord as

$$A(\theta) = 2r \sin \beta \tag{1.19}$$

where r is the radius of the circle. Now the total length of the arc is the same as that of the straight line when all vectors were in phase (i.e., *as*); hence,

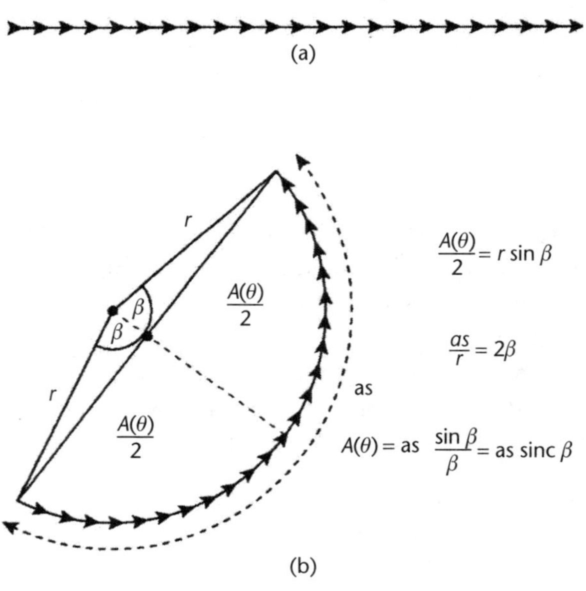

Figure 1.13 Graphical explanation of the sinc function: (a) vectors in phase; and (b) vectors with a progressive phase advance.

$$\frac{as}{t} = 2\beta$$

and thus, substituting for r in (1.19) we have

$$A(\vartheta) = as\frac{\sin\beta}{\beta}$$

Hence the resultant intensity at angle ϑ will be

$$I(\vartheta) = (as)^2\frac{\sin^2\beta}{\beta^2} = I_0\frac{\sin^2\beta}{\beta^2}$$

as in (1.18).

The reason for the ubiquity of this variation in physics can now be seen to be due to the fact that one very often encounters situations where there is a systematically increasing phase difference among a large number of infinitesimal vector quantities: optical interference, electron interference, mass spectrometer energies, particle scattering, and so on.

The principles which lead to the sinc function are all exactly the same, and are those which have just been described.

Let us return now to the intensity diffraction pattern for a slit.

$$I(\vartheta) = I_0\frac{\sin^2\beta}{\beta^2}$$

An important feature of this variation is the scale of the angular divergence. The two minima immediately on either side of the principal maximum (at $\vartheta = 0$) occur when

$$\beta = \frac{1}{2}(ks)\sin\vartheta = \pm\pi$$

giving

$$\sin\vartheta = \pm\frac{\lambda}{s}$$

so that, if ϑ is small, the width of the central maximum is given by

$$\theta_w = 2\vartheta = \pm\frac{2\lambda}{s}$$

Thus, the smaller s is for a given wavelength the more quickly the light energy diverges, and vice versa. This is an important determinant of general behavior in optical systems.

As a second example, consider a sinusoidal variation of amplitude over the aperture. The Fourier transform of a sinusoid consists of one positive and one negative "frequency" equally spaced around the origin. Thus, the diffraction pattern consists of just two lines of intensity equally spaced about the center position of the observing screen (Figure 1.14). Those two lines of intensity could themselves be photographed to provide a "two-slit" aperture plate that would then provide a sinusoidal diffraction (interference?) pattern. This latter pattern will be viewed as an "intensity" pattern, however, not an "amplitude" pattern. Consequently, it will not comprise the original aperture, which must have positive and negative amplitude in order to yield just two lines in its diffraction pattern. Thus, while this example illustrates well the strong relationship that exists between the two functions, it also serves to emphasize that the relationship is between the *amplitude* functions, while the observed diffraction pattern is (in the absence of special arrangements) the *intensity* function.

Finally, we consider one of the most important examples of all: a rectangular-wave aperture amplitude function. The function is shown in Figure 1.15. This is equivalent to a set of narrow slits (i.e., to a diffraction grating). The Fourier transform (and hence the Fraunhofer diffraction pattern) will be a set of discrete lines of intensity, spaced uniformly to accord with the "fundamental" frequency of the aperture function, and enveloped by the Fourier transform of one slit. If the aperture function extended to infinity in each direction then the

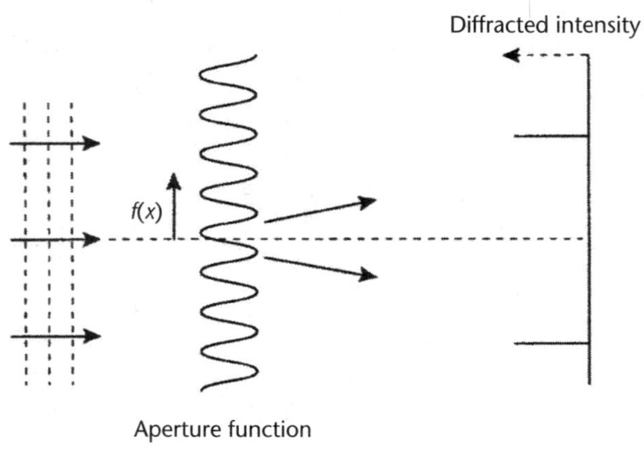

Figure 1.14 Sinusoidal diffracting aperture.

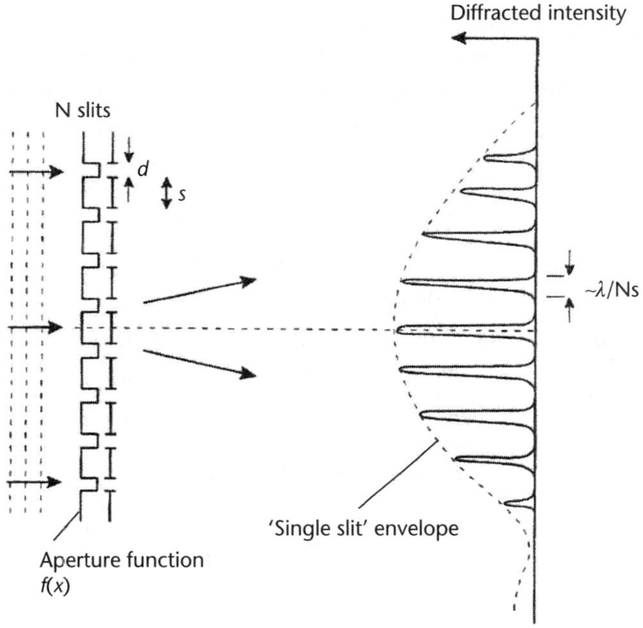

Figure 1.15 Diffraction grating.

individual lines would be infinitely narrow (delta functions), but, since it cannot do so in practice, their width is inversely proportional to the total width of the grating (i.e., the intensity distribution across one line is essentially the Fourier transform of the envelope function for the rectangular wave).

To fix these ideas, consider a grating of N slits, each of width d, and separated by distance s. The diffracted intensity pattern is now given by

$$I(\vartheta) = I_0 \frac{\sin^2 \beta}{\beta^2} \cdot \frac{\sin^2 N\gamma}{\sin^2 \gamma}$$

where

$$\beta = \frac{1}{2}(kd) \sin \vartheta$$

$$\gamma = \frac{1}{2}(ks) \sin \vartheta$$

The pattern is shown in Figure 1.15. Clearly each wavelength present in the light incident on a diffraction grating will produce its own separate diffraction pattern. This fact is used to analyze the spectrum of incident light, and also to select and measure specific component wavelengths. Its ability to perform these tasks is most readily characterized by means of its resolving power, which is defined as

$$\rho = \frac{\lambda}{\delta\lambda}$$

where $\delta\lambda$ is the smallest resolvable wavelength difference. If we take λ to be that wavelength difference which causes the pattern from $\lambda + d\lambda$ to produce a maximum, of order p, which falls on the first minimum of λ at that same order, then we have

$$pN\lambda + \lambda = pN(\lambda + \delta\lambda)$$

and thus

$$\rho = \frac{\lambda}{\delta\lambda} = pN$$

Gratings are ruled either on glass (transmission) or on mirrors (reflection) with ~ 10^5 "lines" (slits) in a ~ 150 mm. The first-order resolving power is thus ~ 10^5.

1.7 Group Velocity

Consider the standard expression for the electric field component of an electromagnetic wave (of arbitrary polarization) propagating in the Oz direction in an optical medium of refractive index n:

$$E = E_0 \exp[i(\omega t - kz)]$$

We know that

$$\frac{\omega}{k} = c = \frac{c_0}{n}$$

and hence may write

$$E = E_0 \exp\left[i\omega\left(t - \frac{nz}{c_0}\right)\right]$$

We may conveniently include both the amplitude attenuation and the phase behavior of the wave in this expression by defining a complex refractive index

$$n = n' - in'' \tag{1.20}$$

so that

$$E = E_0 \exp\left(\frac{-\omega n'' z}{c_0}\right) \exp\left[i\omega\left(t - \frac{n' z}{c_0}\right)\right]$$

The first exponential clearly represents an attenuation factor (real exponent), while the second represents the propagating wave (imaginary exponent).

It has already been noted that refractive index is dependent upon the optical frequency. The physical reason for this is that electromagnetic (primary) waves propagate in a material medium by forcing the elementary atomic/molecular dipoles of the medium into oscillation. These oscillations then radiate their own, secondary, radiation. The extent of this interaction depends upon the relationship between the frequency of the primary, driving, wave and the (fixed) frequencies of the atomic resonances. The atomic oscillators scatter some of the power in the primary wave away from the forward-propagating direction. They also absorb some of it, this component being redistributed as a heating of the material. Both scattering and absorption processes thus lead to attenuation of the primary wave.

Another component of the secondary radiation propagates in the forward direction, combining with the primary wave to produce a resultant forward-propagating wave. However, the phase of the secondary radiation differs from that of the primary wave (as, in general, is always the case for driven oscillators) so that the resultant's phase also differs. This phase change is equivalent to a velocity difference, and this defines the refractive index. The strength of all of these effects is greatest when the frequency of the driving wave coincides with that of an atomic resonance. Hence this also will be the point at which the refractive index is changing most rapidly with optical frequency.

All real sources of light provide their radiation over a range of frequencies. This range is large for an incandescent radiator such as a light bulb, and very small for a gas laser; but it can never be zero. Consequently, in the cases of a medium whose refractive index varies with frequency, different portions of the source spectrum will travel at different velocities and thus will experience different refractive indices. This causes *dispersion* of the light energy, and the medium is thus said to be *optically dispersive*.

The phenomenon has a number of manifestations and practical consequences. One of the best known manifestations is that of the rainbow, where the variation of the refractive index with wavelength in water causes raindrops in the atmosphere to refract the sun's rays through different angles, according to the color of the light, and thus to provide for us a wonderful technicolor display. Another well-known example of dispersion is the experiment performed by Isaac Newton with a glass prism, allowing him to demonstrate quantitatively the different angles of refraction in glass for the spectral colors of which the sun's light is composed.

In the modern idiom of present-day optoelectronics we are rather more concerned with the effect that dispersion has on the information carried by a light beam, especially a guided one; so it is useful to quantify the dispersion effect with this in mind. In order to understand some of these consequences of dispersion, suppose that just two closely spaced frequency components, of equal amplitude, are present in the source spectrum; that is,

$$E = E_0 \cos(\omega t - kz) + E_0 \cos\left(\overline{\omega + \delta\omega}t - \overline{k + \delta k}z\right)$$

where $\delta\omega$, δk are small compared with ω and k, respectively.

Using elementary trigonometry we have

$$E = 2E_0 \cos\frac{1}{2}(\delta\omega t - \delta kz) \cos\left(\overline{\omega + \frac{1}{2}\delta\omega}t - \overline{k + \frac{1}{2}\delta k}z\right)$$

This represents a sinusoidal wave (second cosine factor) whose amplitude is modulated by another sinusoid (first cosine factor) of lower frequency (Figure 1.16). The wave itself travels at a velocity

$$\frac{\omega + \frac{1}{2}\delta\omega}{k + \frac{1}{2}\delta k} \approx \frac{\omega}{k} = c$$

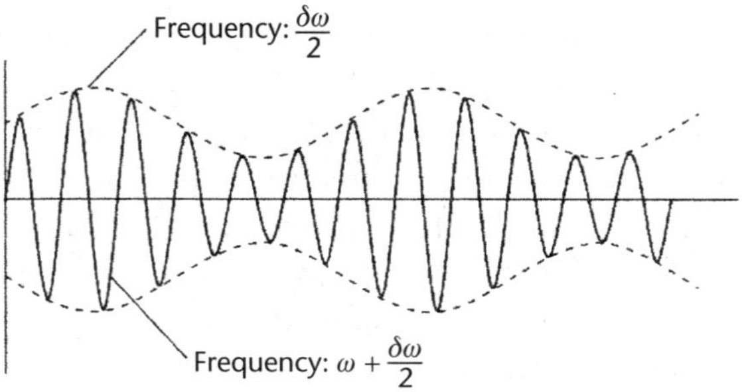

Figure 1.16 Amplitude-modulated wave: sum of two waves of different frequencies.

which is the mean velocity of the two waves. However, the point of maximum amplitude of the wave will always occur when the amplitude modulation has maximum value; that is, when

$$\frac{1}{2}\,\delta\omega t - \frac{1}{2}\,\delta k z = 0$$

so that

$$\frac{\delta\omega}{\delta k} = \frac{z}{t} = c_g$$

and hence, in the limit as $\delta\omega,\ \delta k \to 0$:

$$c_g = \frac{d\omega}{dk} \tag{1.21}$$

where c_g is called the *group* velocity and is the velocity (in this case) with which any given wave maximum progresses. This generalizes to be true for a continuous spread of frequencies, over a small range, such as would be emitted by any practical light source.

Now we also know that

$$\frac{\omega}{k} = c = \frac{c_0}{n}$$

and hence $\omega = (c_0/n)k$ where n is the refractive index of the medium.

In general, n will vary with optical frequency and thus will be a function of k, so that we can differentiate this expression for ω to obtain

$$\frac{d\omega}{dk} = \frac{c_0}{n}\left(1 - \frac{k}{n}\frac{dn}{dk}\right)$$

or, in terms of the wavelength λ:

$$c_g = \frac{d\omega}{dk} = \frac{c_0}{n}\left(1 + \frac{\lambda}{n}\frac{dn}{d\lambda}\right) \tag{1.22}$$

If n does not vary with wavelength, then

$$\frac{dn}{d\lambda} = \frac{dn}{dk} = 0$$

and then

$$\frac{d\omega}{dk} = c_g = \frac{c_0}{n} = c$$

However, if $dn/d\lambda \neq 0$ (i.e., the medium is dispersive) then $c_g \neq c$ and the maximum of the disturbance travels at a different velocity from the "carrier" optical wave. These ideas may readily be generalized to include the complete spectrum of a practical source. Provided that $dn/d\lambda$ is sensibly constant over the spectrum of wavelengths, it follows that a pulse of light from the source will effectively travel undistorted at a velocity of c_g rather than c.

1.8 Emission and Absorption of Light

In considering the processes by which light is emitted and absorbed by atoms, we must again quickly come to terms with the corpuscular or, to use the more modern term, the particulate nature of light.

In classical (i.e., prequantum theory) physics, the atom was held to possess natural resonant frequencies. These corresponded to the electromagnetic wave frequencies that the atom was able to emit when excited into oscillation. Conversely, when light radiation at any of these frequencies fell upon the atom, the atom was able to absorb energy from the radiation in the way of all classical resonant system driving-force interactions.

However, these ideas are incapable of explaining why, in a gas discharge, some frequencies which are emitted by the gas are not also absorbed by it under quiescent conditions; neither can it explain why, in the photoelectric effect (where electrons are ejected from atoms by the interaction with light radiation), the energy with which the electrons are ejected depends not on the intensity of the light, but only on its frequency.

We know that the explanation of those facts is that atoms and molecules can exist only at discrete energy levels. These energy levels may be listed in order of ascending magnitude, $E_0, E_1, E_2, \ldots, E_n$. Under conditions of thermal equilibrium the number of atoms having energy E_i is related to that having energy E_j by the Boltzmann relation:

$$\frac{n_i}{n_j} = \exp\left(-\frac{E_i - E_j}{kT}\right)$$

where k is Boltzmann's constant and T is the absolute temperature. Light can only be absorbed by the atomic system when its frequency ν corresponds to at least one of the values ν_{ji} where

$$\nu_{ji} = E_j - E_i \qquad (j > i)$$

(The symbol v is used now for the frequency rather than $\omega/2\pi$, to emphasize that the light is exhibiting its particulate character.) Here, h is Planck's quantum constant, with value 6.626×10^{-34} joule.seconds.

In this case the interpretation is that one quantum of light, or photon, with energy $h\nu_{ji}$, has been absorbed by the atom, which in consequence has increased in energy from one of its allowed values E_i, to another, E_j. Correspondingly, a photon will be emitted when a downward transition occurs from E_j to E_i, this photon having the same frequency ν_{ji}.

In this context we must think of the light radiation as a stream of photons. If there is a flux of p photons across unit area per unit time then we may write

$$I = ph\nu$$

where I is the light intensity defined in (1.6a). Similarly, any other quantity defined within the wave context also has its counterpart in the particulate context.

In attempting to reconcile the two views, the electromagnetic wave should be regarded as a probability function whose intensity at any point in space defines the probability of finding a photon there. But only in the specialized study of quantum optics are such concepts of real practical significance. For almost all other purposes (including the present one) either the wave representation or the particle representation is appropriate in any given practical situation, without any mutual contradiction.

1.9 Elements of Photodetection

In order to study polarization optics, we can only ever observe optical powers. We analyze polarization phenomena via the effects that polarization elements have on the optical powers. Optical powers are measured as a flux of photons, each photon having an energy $h\nu$, where ν is the optical frequency. The processes which enable light powers to be measured accurately depend directly upon the effects which occur when photons strike matter. In most cases of quantitative measurement the processes rely on the photon to raise an electron to a state where it can be observed directly as an electric current.

We shall not deal in detail with the solid-state physics of photodetection; this is covered in many excellent specialized texts (see, for example, [3]). However, the essentials can be readily appreciated, as follows.

An optical wave arrives at a photodetector as a random stream of particles (photons), obeying (usually) Poisson statistics.

Consider the semiconductor *p-n* junction of Figure 1.17. The physical contact between these two types of semiconductor (i.e., *p* and *n*) leads to a diffusion of majority carriers across the junction in an attempt to equalize their concentrations on either side. The result is to establish an electric field across the junction as a consequence of the charge polarization. Suppose now that a photon is incident upon the region of the semiconductor exposed to this field. If this photon has sufficient energy to create an electron-hole pair, these two new charge carriers will be swept quickly in opposite directions across the junction to give rise to an electric current which can then be measured. The process is assisted by application of an external "reverse bias" electric field. This simple picture of the process enables us to establish two important relationships appropriate to such photodetection devices (which are called photodiodes).

First, for the photon to yield an electron-hole pair its optical frequency (ν) must satisfy $h\nu = E_g$, where E_g is the band-gap energy (the energy required to raise the electron from the valence band to the conduction band) of the material. If ν is too high, however, all the photons will be absorbed in a thin surface layer and the charge pairs will not be collected efficiently by the junction. Thus there is a frequency "responsivity" spectrum for each type of photodiode, which, consequently, must be matched to the spectrum of the light which is to be detected. A typical spectrum for a silicon photodiode is shown in Figure 1.18.

Secondly, suppose that we are seeking to detect a light power of P at an optical frequency ν. This means that $P/h\nu$ photons are arriving every second. Suppose now that a fraction η of these produce electron-hole pairs. Then there are $\eta P/h\nu$ charge carriers of each sign produced every second, so, if all are collected, the observed electric current is given by $i = e\eta P/h\nu$, where e is the electronic charge.

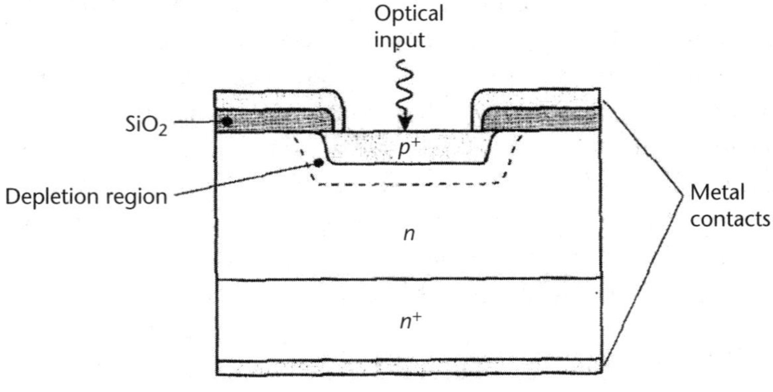

Figure 1.17 Basic structure for a *p-n* junction silicon photodiode.

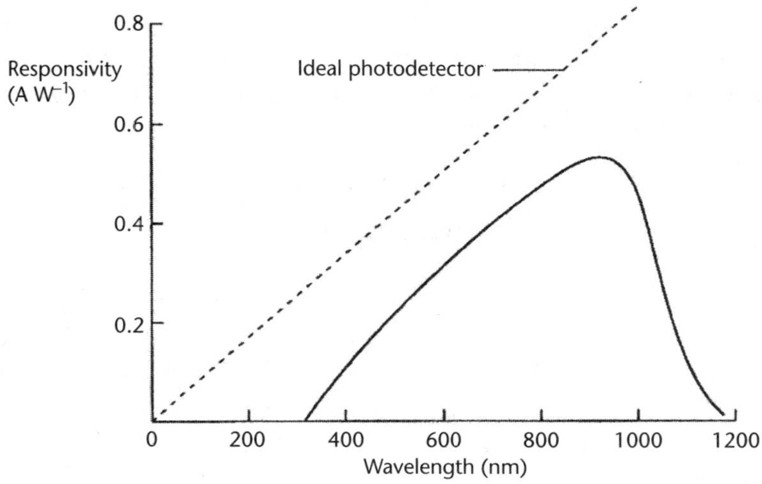

Figure 1.18 Responsivity spectrum for a silicon photodiode.

Thus the current is proportional to the optical power. This means that the electrical power is proportional to the square of the optical power. It is important, therefore, when specifying the signal-to-noise ratio for a detection process, to be sure about whether the ratio is stated in terms of electrical or optical power. (This is a fairly common source of confusion in the specification of detector noise performance.)

Hence optical power is measured via a measurement of the electric current to which it gives rise in a photodiode.

1.10 Conclusions

The wave description of light provides a valuable and powerful analytical tool for the understanding and manipulation of many of its properties. Maxwell's wave equation was an important advance that established the electromagnetic nature of light and pointed the way towards an understanding of many of its interactions with matter.

Using the wave description, we have seen in this chapter how it is possible to explain satisfactorily the phenomena of reflection, refraction, interference, and diffraction. We noted that the light wave is comprised of field vibrations that take place transversely to the propagation direction; we have touched only briefly, however, on the effects that depend upon the particular transverse direction in which this takes place—that is, upon the polarization state of the light.

The polarization properties of light can only be measured via a manipulation of light powers. In order to measure light powers, we need to invoke its corpuscular properties. A flow of corpuscles (photons) gives rise, in a photodetector, to a stream of electrons that is then measured as an electric current.

References

[1] Bleaney, B. I., and B. Bleaney, *Electricity and Magnetism*, Oxford, U.K.: Clarendon Press, 1975.

[2] Born, M., and E. Wolf, *Principles of Optics*, 5th ed., Oxford, U.K.: Pergamon Press, 1975, Section 8.3.2.

[3] Sze, S. M., *Physics of Semiconductor Devices*, New York: John Wiley & Sons, 1981.

Selected Bibliography

Born, M., and E. Wolf, *Principles of Optics*, 5th ed., Oxford, U.K.: Pergamon Press, 1975 (for an excellent rigorous mathematical treatment of classical optics).

Guenther, R., *Modern Optics*, New York: John Wiley and Sons, 1990 (for general wave optics).

Hecht, E., *Optics*, 2nd ed., Reading, MA: Addison-Wesley, 1987, Chapters 9 and 10 (for particularly good treatments of wave interference and diffraction, respectively).

Lipson, S. G., and H. Lipson, *Optical Physics*, Cambridge, U.K.: Cambridge University Press, 1969 (for physical insight into most important wave-optical processes).

2

Optical Waveguiding

2.1 Introduction

In this chapter we continue with the theme of light as a wave motion. The natural tendency for light from a localized source to spread in space, via the phenomenon of diffraction, implies an intrinsic lack of control over the final destination of the optical disturbance. In order to effectively manipulate light at any level of sophistication, such control is clearly required, and this is obtained by means of waveguiding.

Waveguides are physical channels that restrict and configure the physical paths which the light can take, and allow a defined passage between given points in space.

The principles of waveguiding rely on just those wave properties of refraction, interference, polarization and total internal reflection that were established in Chapter 1. In order to make use of these principles for the design and application of practical optical waveguides, it is necessary (as always) to develop an appropriate mathematical description of waveguiding action. This is the task on which we now embark.

2.2 The Slab Waveguide

Consider, first, the symmetrical dielectric structure shown in Figure 2.1. Here we have an infinite (in width and length) dielectric slab of refractive index n_1, sandwiched between two other infinite slabs each of refractive index n_2. This is the easiest arrangement to analyze mathematically, yet it illustrates all the important principles.

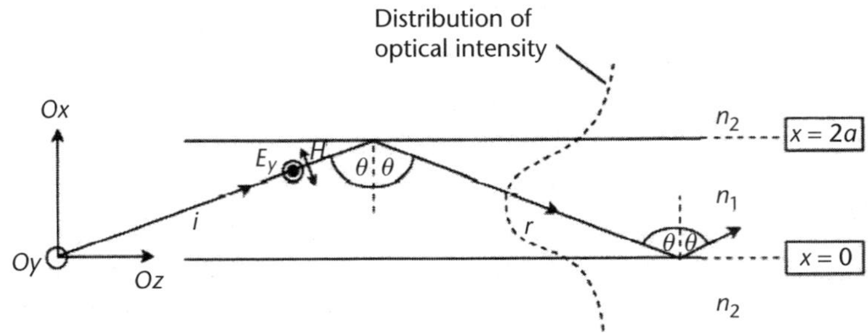

Figure 2.1 Optical slab waveguide.

Using the Cartesian axes defined in the figure, let us consider a light ray (which is, of course, the normal to a propagating plane wavefront) starting at the origin of axes and propagating within the first medium at an angle ϑ with respect to the normal to the boundary between the media. If ϑ is greater than the critical angle (ϑ_c), the light will bounce down the first medium by means of a series of total internal reflections at the boundaries with the other media. Since the wave is thus confined to the first medium, it is said to be *guided* by the structure, which is consequently called a *waveguide*. Let us, firstly, consider guided light which is linearly polarized normal to the plane of incidence. The electric field of the wave represented by ray i (see Figure 2.1) can be written:

$$E_i = E_0 \exp(i\omega t - ikn_1 x \cos\vartheta - ikn_1 z \sin\vartheta)$$

That represented by r, the ray reflected from the first boundary, can be written:

$$E_r = E_0 \exp(i\omega t + ikn_1 x \cos\vartheta - ikn_1 z \sin\vartheta + i\delta_s)$$

where δ_s is the phase change at TIR for this polarization. These two waves will be superimposed on each other and will thus interfere. The interference pattern is obtained by adding them:

$$\begin{aligned} E_T &= E_i + E_r \\ &= E_0 \exp(i\omega t - ikn_1 z \sin\vartheta) \\ &\quad \{\exp(-ikn_1 x \cos\vartheta - i\delta_s/2 + i\delta_s/2) \\ &\quad\quad + \exp(ikn_1 x \cos\vartheta + i\delta_s/2 + i\delta_s/2)\} \\ &= E_0 \exp(i\omega t - ikn_1 z \sin\vartheta + i\delta_s/2) \cdot 2\cos(kn_1 x \cos\vartheta + i\delta_s/2) \end{aligned} \tag{2.1}$$

This is a wave propagating in the direction Oz with wave number $kn_1 \sin\vartheta$, and it is amplitude-modulated in the Ox direction according to

$$\cos\left(kn_1 x \cos\vartheta + \frac{1}{2}\delta_s\right)$$

Now if the wave propagating in the *Oz* direction is to be a stable, symmetrical entity resulting from a self-reproducing interference pattern, the intensity of the wave must be the same at each of the two boundaries. This requires that it is the same for $x = 0$ as for $x = 2a$; that is,

$$\cos^2\left(\frac{1}{2}\delta_s\right) = \cos^2\left(kn_1 2a \cos\vartheta + \frac{1}{2}\delta_s\right) \tag{2.2}$$

The general solution on this equation is

$$\frac{1}{2}\delta_s = m\pi \pm \left(2akn_1 \cos\vartheta + \frac{1}{2}\delta_s\right)$$

where m is any integer (positive or negative).

Hence, either

$$2akn_1 \cos\vartheta + \delta_s = m\pi \qquad (-) \tag{2.3a}$$

or

$$2akn_1 \cos\vartheta = -m\pi \qquad (+) \tag{2.3b}$$

However, there is another condition to impose. If the interference pattern is to self-reproduce in a consistent way as it propagates down the guide, the phase change experienced by a ray executing one complete "bounce" down the guide must be an integer times 2π. In order to progress down the guide indefinitely, the waves from the successive boundaries must interfere constructively, forming what is essentially a continuous, stable, interference pattern down the guiding channel. If the interference is not fully constructive, the wave will eventually "self-destruct," owing to the out-of-phase cancellations (although, clearly, if the phasings are almost correct, the wave might persist for a considerable distance, attenuating only slowly). The wavefronts resulting from ray reflections at all points along the guide can only be in phase provided that

$$2akn_1 \cos\vartheta + \delta_s = m\pi$$

which corresponds to (2.3a). Equation (2.3b) does not satisfy the condition on wavefronts and is, therefore, invalid. Equation (2.3a) is sometimes known as the *transverse resonance condition* since it corresponds essentially to the condition that, when resolving the wave vector into directions transverse and parallel to the guide axis, the transverse component has just one half cycle, or an integer multiple thereof ($m\pi$), fitting into the guide width. This is a "resonance" in the sense that a string stretched between two points resonates, when plucked, at frequencies which are conditioned in just the same way.

Now since δ_s depends only on ϑ (see Fresnel's equations in Section 1.3), it follows that the condition

$$2akn_1 \cos \vartheta + \delta_s = m\pi$$

is a condition on ϑ. The condition tells us that ϑ can have only certain discrete values if the interference pattern is to remain constant in form along the length of the fiber. Each form of interference pattern is, therefore, characterized by a particular value of m, which then provides a corresponding value for ϑ. The allowed interference patterns are called the *modes* of the waveguide, for they are determined by the properties (geometrical and physical) of the guide.

Now the wavenumber, k ($= 2\pi/\lambda$), for the free space propagation of the wave has suffered a number of modifications. First, the wavelength of the light is smaller in the medium than in free space (the frequency remains the same, but the velocity is reduced by a factor $n_{1,2}$), so we can conveniently define

$$\beta_1 = n_1 k$$

$$\beta_2 = n_2 k$$

as the wavenumbers in the guiding and outer slabs, respectively. Secondly, however, if we choose to interpret (2.1) as one describing a wave propagating in the Oz direction with amplitude modulated in the Ox direction, it is convenient to resolve the wavenumber in the guiding medium into components along Oz and Ox; that is,
along Oz:

$$\beta = n_1 k \sin \vartheta \tag{2.4a}$$

along Ox:

$$q = n_1 k \cos \vartheta \tag{2.4b}$$

Of these two components, β is clearly the more important, since it is the effective wavenumber for the propagation down the guide. In fact, (2.1) can now be written:

$$E_T = 2E_0 \cos\left(qx + \frac{1}{2}\delta_s\right) \exp i\left(\omega t - \beta z + \frac{1}{2}\delta_s\right)$$

What can be said about the velocity of the wave down the guide? Clearly the phase velocity is given by

$$c_p = \frac{\omega}{\beta}$$

However, the velocity with which optical energy propagates down the guide is given by the group velocity, which, in this case, is given by (see Section 1.7)

$$c_g = \frac{d\omega}{d\beta}$$

Clearly, to evaluate this, we require the dependence of ω upon β. To obtain this, let us start with (2.4a); that is,

$$\beta = n_1 k \sin \vartheta$$

The first thing to note is that, for all real ϑ, this requires

$$\beta \leq n_1 k$$

Also, since the TIR condition requires that

$$\sin \vartheta \geq \frac{n_2}{n_1}$$

it follows that

$$\beta = n_1 k \sin \vartheta \geq n_2 k$$

Hence, we have

$$n_1 k \geq \beta \geq n_2 k$$

or

$$\beta_1 \geq \beta \geq \beta_2$$

In other words, the wavenumber describing the propagation along the guide axis always lies between the wavenumbers for the guiding medium (β_1) and the outer medium (β_2). This we might have expected from the physics, since the propagation lies partly in the guide and partly in the outer medium (evanescent wave). We shall be returning to this point later.

Remember that our present concern is about how β varies with ω between these two limits, so how else does (2.4a) help?

Clearly, the relation

$$k = \frac{\omega}{c_0}$$

where c_0 is the free space velocity, gives one dependence of β on ω, but what about sin ϑ? For a given value of m (i.e., a given mode) the transverse resonant condition (2.3a) provides the dependence of ϑ on k. However, this is quite complex since, as we know, δ_s is a quite complex function of ϑ. Hence, in order to proceed further, this dependence must be considered.

The expressions for the phase changes which occur under TIR at a given angle were derived in Section 1.4 and are restated here:

$$\tan \frac{1}{2} \delta_s = \frac{\left(n_1^2 \sin^2 \vartheta - n_2^2\right)^{1/2}}{n_1 \cos \vartheta}$$

for the case where the electric field is perpendicular to the plane of incidence, and

$$\tan \frac{1}{2} \delta_p = \frac{n_1\left(n_1^2 \sin^2 \vartheta - n_2^2\right)^{1/2}}{n_2^2 \cos \vartheta}$$

for the case where it lies in the plane of incidence.

Note also that

$$\tan \frac{1}{2} \delta_p = \frac{n_1^2}{n_2^2} \tan \frac{1}{2} \delta_s$$

Finally, let us define, for convenience, a parameter, p, where

$$p^2 = \beta^2 - n_2^2 k^2 = k^2\left(n_1^2 \sin^2 \vartheta - n_2^2\right) \qquad (2.5)$$

The physical significance of p will soon become clear.

We now discover that we can cast our *transverse resonance* condition (2.3a) into the form

$$\tan\left(aq - \frac{1}{2}m\pi\right) = \frac{p}{q} \qquad (E_{\perp}) \tag{2.6a}$$

for the perpendicular polarization and

$$\tan\left(aq - \frac{1}{2}m\pi\right) = \frac{n_1^2}{n_2^2}\frac{p}{q} \qquad (E_{\parallel}) \tag{2.6b}$$

for the parallel polarization.

The conventional waveguide notation designates these two cases as *transverse electric* (TE) for $E_{\perp}$ and *transverse magnetic* (TM) for $E_{\parallel}$. The terms refer, of course, to the direction of the stated fields with respect to the plane of incidence of the ray.

We can use (2.6) to characterize the modes for any given slab geometry. The solutions of the equations can be separated into odd and even types according to whether m is odd or even. For odd m we have (from trigonometrical identities)

$$\tan\left(aq - \frac{1}{2}m_{\text{odd}}\pi\right) = -\cot aq \tag{2.7a}$$

and for even m we have

$$\tan\left(aq - \frac{1}{2}m_{\text{even}}\pi\right) = \tan aq \tag{2.7b}$$

Taking m to be even, we may then write (2.6a), for example, in the form:

$$aq \tan aq = ap \qquad (E_{\perp}) \tag{2.8}$$

Now from the definitions of p and q it is clear that

$$a^2p^2 + a^2q^2 = a^2k^2\left(n_1^2 - n_2^2\right) \tag{2.9}$$

Taking rectangular axes ap, aq, this latter relation between p and q translates into a circle of radius $ak\left(n_1^2 - n_2^2\right)^{1/2}$ (see Figure 2.2). If, on the same axes, we also plot the function $aq \tan aq$, then (2.8) is satisfied at all points of intersection between the two functions (Figure 2.2). (A similar set of solutions clearly can

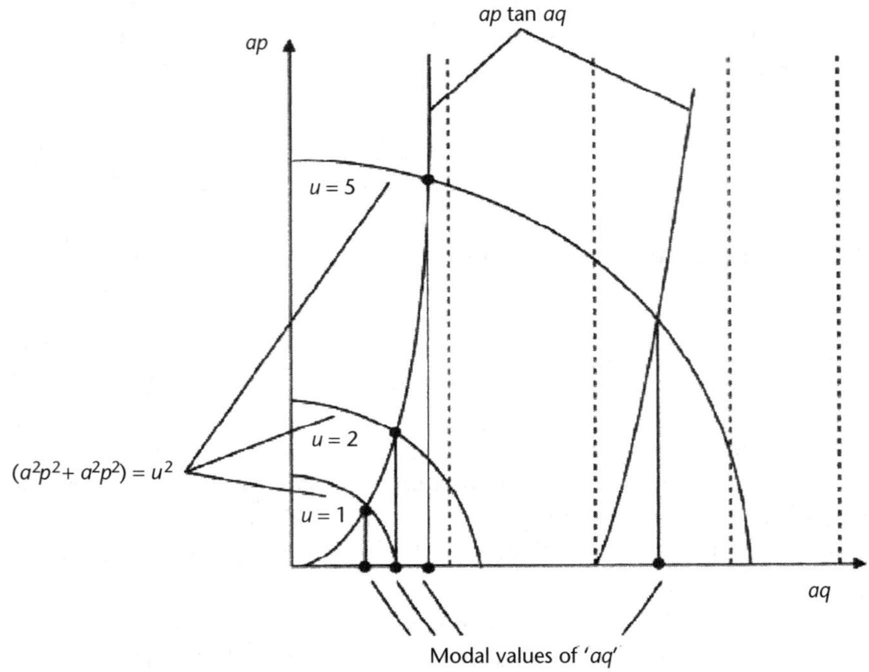

Figure 2.2 Graphical solution of the modal equation for the slab waveguide.

be found for odd m.) These points, therefore, provide the values of ϑ that correspond to the allowed modes of the guide. Having determined a value for ϑ for a given k, β can be determined from

$$\beta = n_1 k \sin \vartheta$$

and hence β can be determined as a function of k (for a given m) for the TE modes.

Now, finally, with

$$k = \frac{\omega}{c}$$

we have the relationship between β and ω which we have been seeking. For obvious reasons these are called dispersion curves, and are important determinants of waveguide behavior. They are drawn either as β versus k or as ω versus β. The three lowest order modes for a typical slab waveguide are shown in Figure 2.3(a) using the latter representation. Clearly, this is the more convenient form for determining the group velocity $d\omega/d\beta$ by simple differentiation [Figure 2.3(b)].

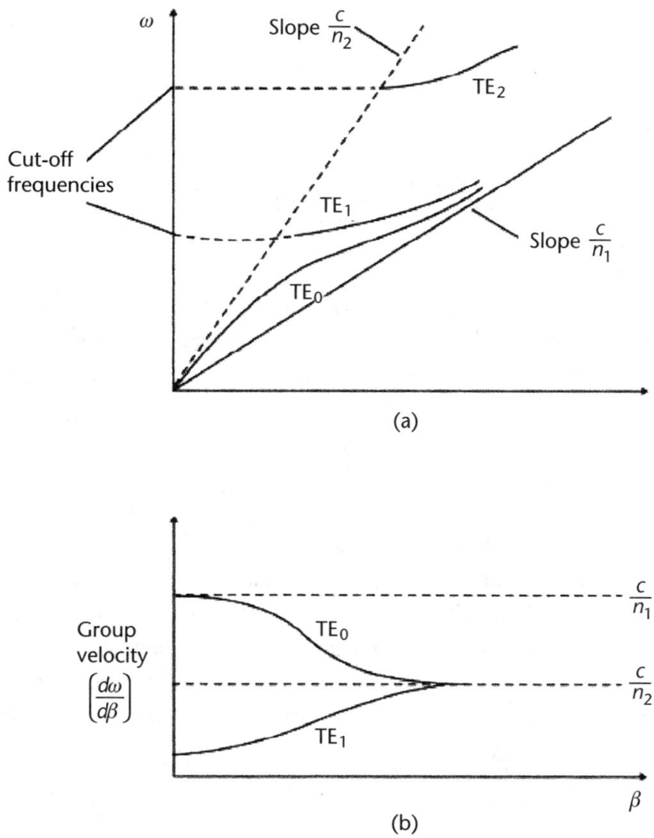

Figure 2.3 Dispersion and group velocity for the slab waveguide: (a) dispersion diagram for the slab waveguide; and (b) variation of group velocity with wavenumber.

A final point of great importance should be made. As k decreases, so the quantity

$$a^2p^2 + a^2q^2 = a^2k^2\left(n_1^2 - n_2^2\right)$$

decreases and the various modes are sequentially "cut off" as the circle (Figure 2.2) reduces in radius. This is also apparent in Figure 2.3(a) since a reduction in k corresponds, of course, to a reduction in ω. Clearly the number of possible modes depends upon the waveguide parameters a, n_1, and n_2. However, it is also clear that there will always be at least one mode, since the circle will always intersect the tan curve at one point, even for a vanishingly small circle radius. If there is only one solution, for any value of m, then Figure 2.2 shows that the radius of the circle must be less than $\frac{1}{2}\pi$; that is,

$$ak\left(n_1^2 - n_2^2\right)^{1/2} < \frac{1}{2}\pi$$

or

$$\frac{2\pi a}{\lambda}\left(n_1^2 - n_2^2\right)^{1/2} < 1.57 \tag{2.10a}$$

This quantity is another important waveguide parameter, for this and many other reasons. It is given the symbol V and is called the *normalized frequency*, or, quite often, simply the *V number*. Thus,

$$V = \frac{2\pi a}{\lambda}\left(n_1^2 - n_2^2\right)^{1/2}$$

Equation (2.10) is thus the single-mode condition for this symmetrical slab waveguide. It represents an important case, since the existence of just one mode in a waveguide simplifies considerably the behavior of radiation within it, and thus facilitates its use in, for example, the transmission of information along it. Physically, (2.10) is stating the condition under which it is impossible for constructive interference to occur for any ray other than that which (almost) travels along the guide axis.

Clearly, a very similar analysis can be performed for the TM modes, using (2.6b).

Look again now at Figure 2.1. It is shown that there are waves traveling in the outer media with amplitudes falling off the farther we go from the central channel, for the intensity of the interference pattern is nonzero in the cladding. This matter was dealt with in Section 1.4, where it was seen that this was a direct result of the necessity for fields (and their derivatives) to be continuous across the media boundaries. We know from (2.1) that the field amplitude in the central channel varies as

$$E_x = 2E_0 \cos\left(kn_1 x \cos\vartheta + \frac{1}{2}\delta_s\right)$$

How does the field in the outer slabs vary? The answer to this question was given in Section 1.4 when dealing with the TIR phenomenon. It was seen there that the evanescent field in the second medium, when TIR occurred, fell off in amplitude according to

$$E_x = E_a \exp\left(-\frac{2\pi x}{\lambda_2}\sinh\gamma\right); \qquad x > a$$

where

E_a is the value of the field at the boundary; that is,

$$E_a = 2E_0 \cos\left(kn_1 a \cos\vartheta + \frac{1}{2}\delta_s\right)$$

λ_2 is the wavelength in the second medium, and is equal to λ/n_2;

$2\pi \frac{\sinh\gamma}{\lambda_2} = k\left(n_1^2 \sin^2\vartheta - n_2^2\right)^{1/2}$, and this can now be identified with p from (2.5).

Hence,

$$E_x = E_a \exp(-px); \qquad x > a \tag{2.10b}$$

and we see that p is just the exponential decay constant for the amplitude of the evanescent wave (Figure 2.4) and, from (2.5), we note that $p \sim 0.1k$. (It is a fact of any physical analysis that all parameters of mathematical importance will always have a simple physical meaning.)

So the evanescent waves are waves that propagate in the outer media parallel with the boundary but with amplitude falling off exponentially with distance from the boundary.

These evanescent waves are very important. First, if the total propagation is not to be disturbed by the external environment, the thickness of each outer

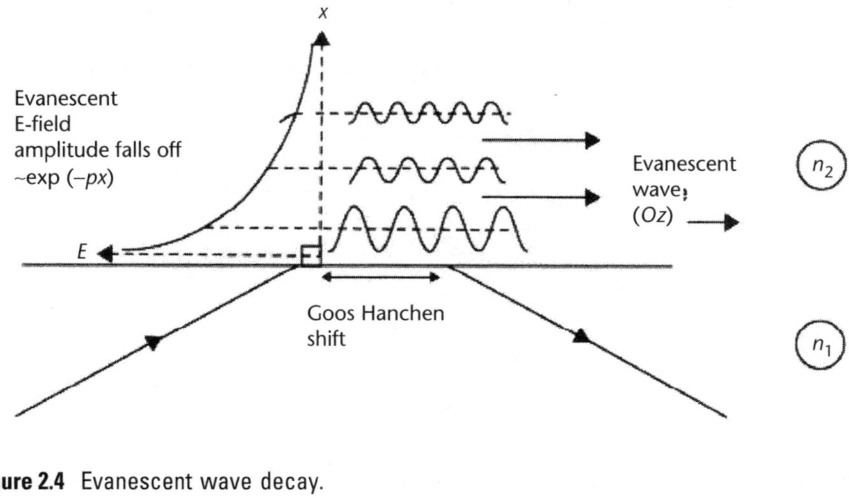

Figure 2.4 Evanescent wave decay.

slab must be great enough for the evanescent wave to have negligible amplitude at its outer boundary: the wave falls off as ~ $\exp(-x/\lambda)$, so at $x \sim 20\lambda$ it normally will be quite negligible (~ 10^{-9}). At optical wavelengths, then, the slabs should have a thickness ≥ 20 μm.

Secondly, since energy is traveling (in the *Oz* direction!) in the outer media, the evanescent wave properties will influence the core propagation, in respect, for example, of loss and dispersion. We shall consider these aspects in more detail in Section 2.6.2.

2.3 Integrated Optics

Planar waveguides find interesting application in integrated optics. In this, waves are guided by planar channels and are processed in a variety of ways. An example is shown in Figure 2.5. This is an electro-optic modulator, a device that modifies the power of light passing through it proportionally to an applied electric field. However, the electric field is acting on a waveguide which, in this case, is a channel (such as we have just been considering) surrounded by *outer slabs* called here a *substrate*. The electric field is imposed by means of the two substrate electrodes, and the interaction path is under close control, as a result of the waveguiding. The material of which both the substrate and the waveguide are made should, in this case, clearly be an electro-optic material, such as lithium tantalate ($LiTaO_3$). The central waveguiding channel may be constructed by diffusing ions into it (under careful control); an example of a suitable ion is niobium (Nb), which will thus increase the refractive index of the "diffused" region and allow total internal reflection to occur at its boundaries with the "raw" $LiTaO_3$. Many other functions are possible using suitable materials, geometries, and field influences. It is possible to fabricate couplers, amplifiers, polarizers, filters, and so on, all within a planar "integrated" geometry.

One of the advantages of this integrated optics technology is that the structures can be produced to high manufactured tolerance by mass-production

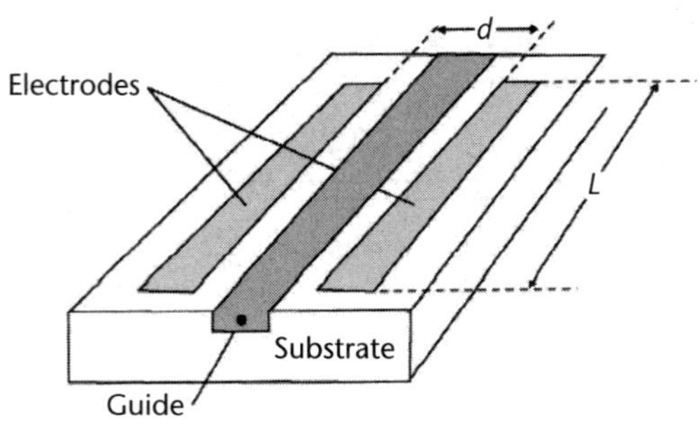

Figure 2.5 An integrated electro-optic phase modulator.

methods, allowing them to be produced cheaply if required numbers are large, as is likely to be the case in optical telecommunications, for example.

A potentially very powerful development is that of the Optoelectronic Integrated Circuit (OEIC) which combines optical waveguide functions with electronic ones such as optical source control, photodetection and signal processing, again on a single, planar, readily-manufacturable chip.

Note, finally, that in Figure 2.5 the upper slab (air) has a different refractive index from the lower one (substrate). This is thus an example of an asymmetrical planar waveguide, the analysis of which is more complex than the symmetrical one that we have considered. However, the basic principles are the same; the mathematics is just more cumbersome, and is covered in many other texts (see, for example, [1]).

2.4 Cylindrical Waveguides

Let us now consider the cylindrical dielectric structure shown in Figure 2.6. This is the geometry of the optical fiber, the central region being known as the *core* and the outer region as the *cladding*. In this case the same basic principles apply as for the dielectric slab, but the circular, rather than planar, symmetry complicates the mathematics. We use, for convenience, cylindrical coordinates (r, φ, z) as defined in Figure 2.6. This allows us to cast Maxwell's wave equation (see Appendix A) for the dielectric structure into the form:

$$\nabla^2 E = \frac{1}{r}\frac{\partial}{\partial r}\left(r\frac{\partial E}{\partial r}\right) + \frac{1}{r^2}\frac{\partial^2 E}{\partial p^2} + \frac{\partial^2 E}{\partial z^2} = \mu\epsilon\frac{\partial^2 E}{\partial t^2} \tag{2.11}$$

If we try a solution for E in which all variables are separable, we write

$$E = E_r(r)E_\varphi(\varphi)E_z(z)E_t(t)$$

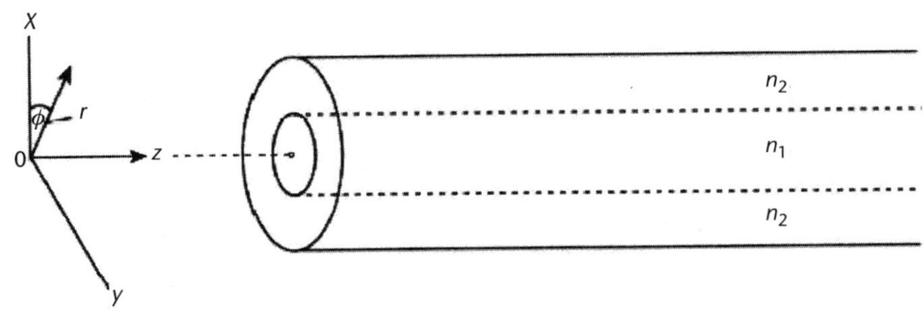

Figure 2.6 Cylindrical waveguide geometry.

and can immediately, from the known physics, take it that

$$E_z(z)E_t(t) = \exp[i(\beta z - \omega t)]$$

In other words, the wave is progressing along the axis of the cylinder with wavenumber β and with angular frequency ω. It follows, of course, that its (phase) velocity of progression along the axis is given by

$$c_p = \frac{\omega}{\beta}$$

By substitution of these expressions into the wave (2.11), we may rewrite it in the form:

$$\frac{\partial}{\partial r}\left(r\frac{\partial(E_r E_\varphi)}{\partial r}\right) + \frac{1}{r^2}\frac{\partial^2(E_r E_\varphi)}{\partial \varphi^2} - \beta^2 E_r E_\varphi + \mu\epsilon\omega^2 E_r E_\varphi = 0$$

Now if we suggest a periodic function for E_φ of the form:

$$E_\varphi = \exp(\pm il\varphi)$$

where l is an integer, we can further reduce the equation to

$$\frac{\partial^2 E_r}{\partial r^2} + \frac{1}{r}\frac{\partial E_r}{\partial r} + \left(n^2k^2 - \beta^2 - \frac{l^2}{r^2}\right)E_r = 0$$

This is a form of Bessel's equation, and its solutions are Bessel functions (see any advanced mathematical text, e.g., [2]). If we use the same substitutions as for the previous planar case—that is,

$$n_1^2k^2 - \beta^2 = q^2$$

$$\beta^2 - n_2^2k^2 = p^2$$

we find for $r \le a$ (core)

$$\frac{\partial^2 E_r}{\partial r^2} + \frac{1}{r}\frac{\partial E_r}{\partial r} + \left(q^2 - \frac{l^2}{r^2}\right)E_r = 0$$

and for $r > a$ (cladding)

$$\frac{\partial^2 E_r}{\partial r^2} + \frac{1}{r}\frac{\partial E_r}{\partial r} + \left(p^2 - \frac{l^2}{r^2}\right)E_r = 0$$

Solutions of these equations are [see Figure 2.7(a)]

$$E_r = E_c J_l(qr); \qquad r \leq a$$

$$E_r = E_{cl} K_l(pr); \qquad r > a$$

where J_l is a Bessel function of the first kind and K_l is a modified Bessel function of the second kind (sometimes known as a modified Hankel function). The

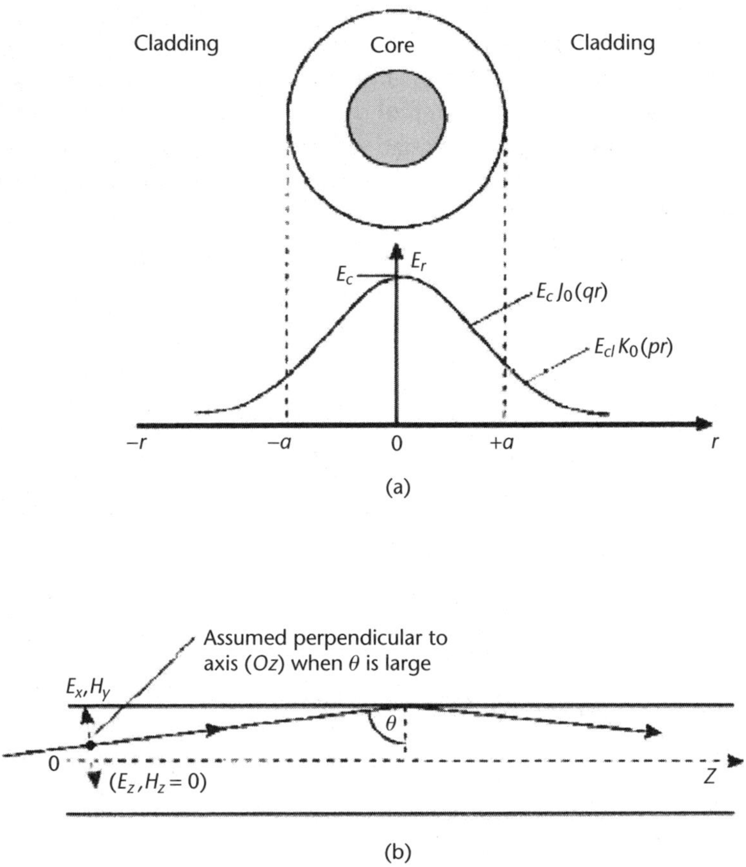

Figure 2.7 (a) Lowest order solution of the cylindrical waveguide equation ($t = 0$). (b) The geometry of the weakly guiding approximation.

two functions clearly must be continuous at $r = a$, and we have for our full "trial" solution in the core

$$E = E_c J_l(qr)\exp(\pm il\varphi)\exp i(\beta z - \omega t)$$

and a similar one for the cladding

$$E = E_{cl} J_l(pr)\exp(\pm il\varphi)\exp i(\beta z - \omega t)$$

Again we can determine the allowable values for p, q, and β by imposing the boundary conditions at $r = a$ [3]. The result is a relationship which provides the β versus k, or *dispersion* curves, shown in Figure 2.8. The mathematical manipulations are tedious, but are somewhat eased by using the so-called *weakly guiding* approximation. This makes use of the fact that if $n_1 \sim n_2$, then the ray's angle of incidence on the boundary must be very large, if TIR is to occur. The ray must bounce down the core almost at grazing incidence. This means that the wave is very nearly a transverse wave, with very small z components of electric and magnetic fields. By neglecting the longitudinal components H_z, E_z, a considerable simplification of the mathematics results [Figure 2.7(b)]. Since the wave is, to a first approximation, transverse, it can be resolved conveniently into two linearly polarized components, just as for free space propagation. The modes are thus dubbed *linearly polarized* (LP) modes, and the notation that describes the profile's intensity distribution is the *LP notation*.

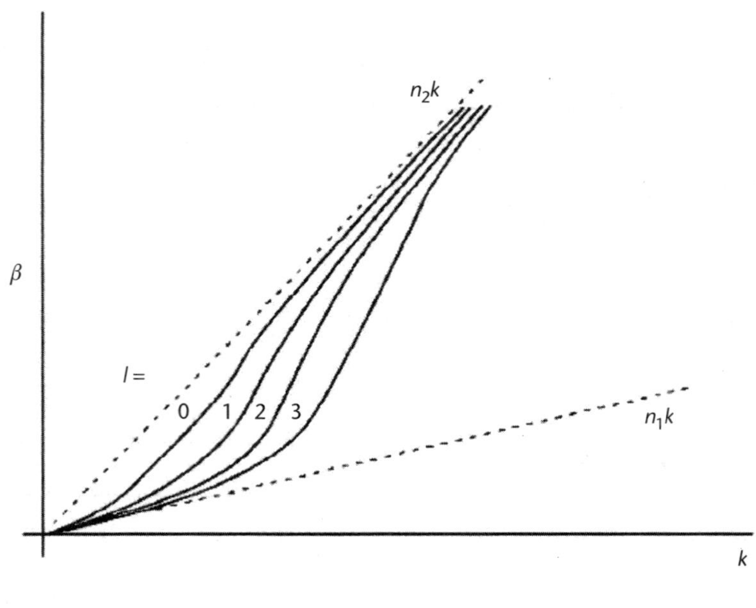

Figure 2.8 Dispersion curves for the cylindrical waveguide.

2.5 Optical Fibers

The cylindrical geometry relates directly, of course, to the optical fiber. The latter has just the geometry we have been considering and, for a typical fiber,

$$\frac{n_1 - n_2}{n_1} \sim 0.01$$

so that the weakly guiding approximation is valid. Some of the low-order LP modes of intensity distribution are shown in Figure 2.9, together with their polarizations, and values for the azimuthal integer, 1. There are, then, two possible linearly polarized optical fiber modes. For the cylindrical geometry the single-mode condition is [analogously to (2.10) for the planar case]

$$V = \frac{2\pi a}{\lambda}\left(n_1^2 - n_2^2\right)^{1/2} < 2.405$$

The number 2.405 derives from the value of the argument for which the lowest order Bessel function, J_0, has its first zero (see Figure 2.10). Some important practical features of optical-fiber design can be appreciated from the above two equations.

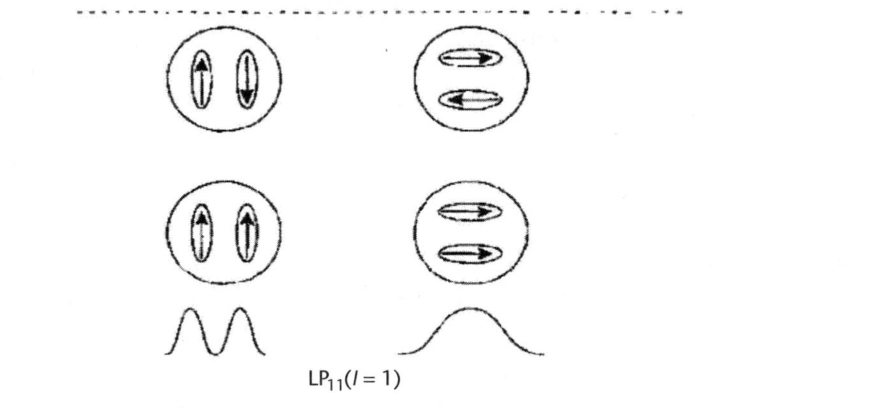

Figure 2.9 Some low order modes for the cylindrical waveguide (with weakly guiding labels).

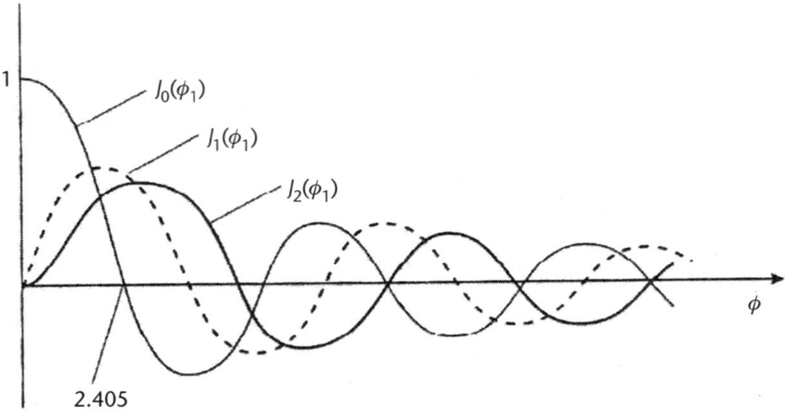

Figure 2.10 Plot of Bessel functions of the first kind.

First, the material of which an optical fiber is made must be transparent to optical wavelengths. Glass, which consists largely of fused silica (SiO_2) usually is chosen, for reasons which will be elaborated upon in Section 2.6.1.

Silica has a refractive index of 1.47 at optical wavelengths. This can be modified by doping to provide for the differing core and cladding values, so that we can readily satisfy the above two equations for single-mode operation at an optical wavelength of, say, 1 μm, by using a core radius, a, of 2 to 3 μm. Typical core diameters for single-mode fibers are thus ~ 5 μm, while those for multimode fibers can range up to ~ 50 μm. Cladding diameters are standardized, for both types of fiber, at 125 μm, thus ensuring that the evanescent field is negligible (for both types) at the outer boundary's interface with the outside world. Further important features can best be appreciated by reversion to geometrical (ray) optics.

As an example, let us consider, first, the problem of launching light into the fiber. Referring to Figure 2.11(a), we have for a ray incident on the front face of the fiber at angle ϑ_0, and with refracted angle ϑ_1:

$$n_0 \sin \vartheta_0 = n_1 \sin \vartheta_1$$

where n_0 and n_1 are the refractive indices of air and the fiber core material, respectively. If the angle at which the ray then strikes the core/cladding boundary is ϑ_T, then, for TIR, we must have: $\sin \vartheta_T > n_2/n_1$ where n_2 is the cladding index.

Since $\vartheta_T = \frac{1}{2}\pi - \vartheta_1$ the inequality is equivalent to

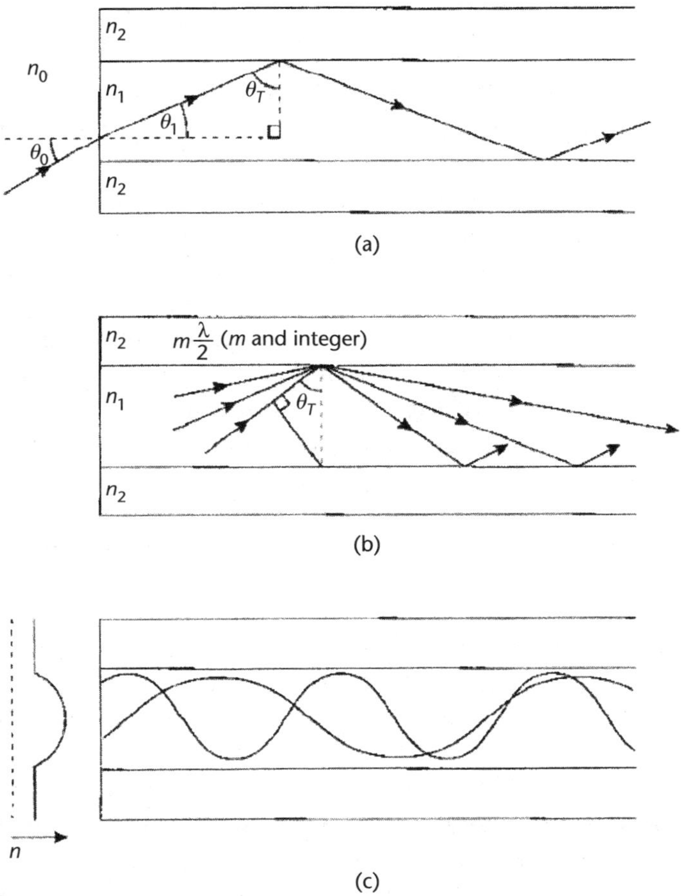

Figure 2.11 Ray propagations in optical fibers: (a) acceptance angle for an optical fiber; (b) ray representations of fiber modes; and (c) graded index ray paths.

$$\cos\vartheta_1 > \frac{n_2}{n_1}$$

so from the Snell's law expression above,

$$\cos\vartheta_1 = \left(1 - \frac{n_0^2 \sin^2\vartheta_0}{n_1^2}\right)^{1/2}$$

or

$$n_0 \sin\vartheta_0 < \left(n_1^2 - n_2^2\right)^{1/2}$$

The quantity on the RHS of this inequality is known as the numerical aperture (NA) of the fiber. It is a specification of the *acceptance* cone of light, this being a cone of apex half-angle ϑ_0. Clearly, a large refractive index difference between core and cladding is necessary for a large acceptance angle; for a typical fiber, $\vartheta_0 \sim 10°$.

The discrete values of reflection angle which are allowed by the transverse resonance condition (within the TIR condition) can be represented by the ray propagations shown in Figure 2.11(b). This makes clear that for a large number of allowable rays (i.e., modes) the TIR angle should be large, implying a large NA. It is also clear, however, geometrically, that the rays will progress down the guide at velocities which depend on their angles of reflection: the smaller the angle, the smaller the velocity. This leads to large *modal dispersion* at large NA since, if the launched light energy is distributed among many modes, the differing velocities will lead to varying times of arrival of the energy components at the far end of the fiber. This is undesirable in, for example, communications applications, since it will lead to a limitation on the communications bandwidth. In a digital system, a pulse cannot be allowed to spread into the pulses before or after it. For greatest bandwidth only one mode should be allowed, and this requires a small NA. Thus a balance must be struck between good signal level (large NA) and large signal bandwidth (small NA).

A fiber design which attempts to attain a better-balanced position between these is shown in Figure 2.11(c). This fiber is known as graded-index (GI) fiber and it possesses a core refractive index profile that falls off parabolically (approximately) from its peak value on the axis. This profile constitutes, effectively, a continuous convex lens, which allows large acceptance angle while limiting the number of allowable modes to a relatively small value. GI fiber is used widely in short and medium distance communications systems. For trunk systems single-mode fiber is invariably used, however. This ensures that the modal dispersion is entirely absent, thus removing this limitation on bandwidth. Single-mode fiber possesses a communications bandwidth that is an order of magnitude greater than that of multimode fiber. However, it is not without its problems. It is time now to deal with the communications application for optical fiber in a more coherent fashion.

2.6 Optical Fibers for Communications

The most important application area for optical fibers (and, arguably, for the whole of photonics) at the present time is that of optical communications.

The basic arrangement for an optical-fiber communications system is shown in Figure 2.12. A laser source provides light that is modulated by the information required to be transmitted, the information being in the form of

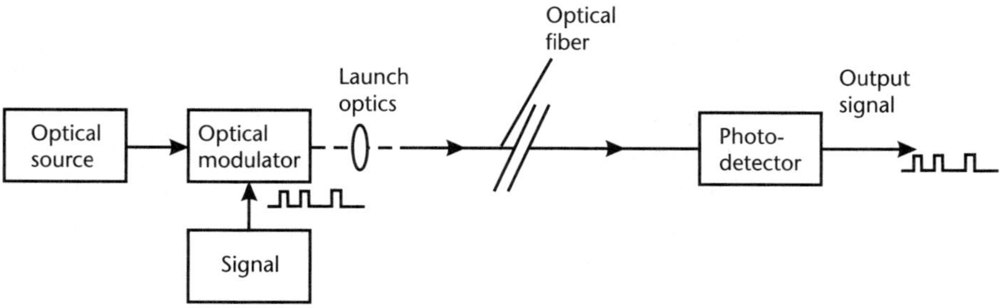

Figure 2.12 Schematic for an optical fiber communications system.

an electrical signal that is applied to an optical modulator. This light is then launched into an optical fiber that guides it to its destination. At the destination the light emerges from the fiber and falls onto a photodetector that converts it into an electrical signal. This electrical signal will be a close reproduction of that which was used to modulate the laser source: the closer it is, the better is the communications link.

The primary advantage of such an optical arrangement is the enormous communications bandwidth that it offers, for bandwidth is equivalent to information-carrying capacity. The reason for this is that the frequency of the light is so much greater than that of the more conventional carrier signals such as radio and microwave transmissions: light in the visible range has a frequency of 5×10^{14} Hz compared with microwaves at $\sim 10^{10}$ Hz and radio at $\sim 10^{8}$ Hz. The higher the frequency of the carrier, the smaller the relative effect of a given modulation bandwidth, for any modulation signal will spread the carrier signal over a band at least equal to the modulation bandwidth. Hence, a 1-GHz (10^9 Hz) modulation bandwidth will spread a microwave carrier by 10%, but an optical carrier by only 0.0002% (using the above figures). The smaller perturbation of the carrier frequency means that the properties of all the components in the communications system are substantially constant over the transmission bandwidth, and this applies especially to the transmission medium. For, if the medium acts differently for different frequencies over the modulation bandwidth, the information becomes distorted, and the communications link performance degrades. Hence, at optical frequencies, using optical-fiber waveguides, very high-bandwidth systems can be expected.

Most long-distance communication systems presently are digital, which means that the signal information exists in the form of a series of pulses that encode the information as a series of "yes" or "no" answers to the question "is a pulse present in a particular time slot?" The advantage of this is that the detection system has only to answer "yes" or "no" to this simple question and does not have to decide on the precise level of the signal over a range, as is the case for analog systems. Digital systems are thus very robust in terms of signal level, the only requirement being that the level should be above a certain threshold, but they do require more bandwidth than analog systems. Optical-fiber communications systems readily provide this. However, even optical fibers both attenuate and distort the transmitted signals to some extent. It is necessary to understand the processes that lead to attenuation and distortion in fibers in order to get the best from them for communications purposes. These are the subjects for the next two sections.

2.6.1 Optical-Fiber Attenuation

The attenuation in silica is due to two primary mechanisms. The first is that of scattering from the small ($<\lambda$) inhomogeneities of the (amorphous) structure and composition in the glass. This is known as Rayleigh scattering and it has the effect of removing light from the guiding region with an efficiency which varies as $1/\lambda^4$. The second is absorption by the atoms and molecules of which the material is composed. The extent of this depends on the relationship between the propagating light frequency and the various natural resonant frequencies characteristic of these atoms and molecules. Close to a resonance, the light will be absorbed and redistributed as a heating of the material, hence being lost from the propagation. There will also be a variation in light velocity as a result of these atomic interactions, and this leads to the phenomenon of dispersion discussed in Section 1.7.

Most high-grade optical fibers are fabricated from amorphous silica by drawing a thin fiber strand from a melt (Figure 2.13). The block of the material that is melted is called the *preform* and it is carefully constructed to have the required scaled-up geometry of the fiber. The core is given greater refractive index than the cladding by doping with materials such as beryllium or germanium. The geometry is preserved in the drawing process. Many tens of kilometers of fiber can be drawn from a single preform. (For details of fiber fabrication processes see, for example, [4].)

A schematic of the absorption spectrum of silica is shown in Figure 2.14. The two peaks shown are in fact harmonics (overtones) of a fundamental vibration at 2.8 μm which is due to the stretching of the O-H bond. These peaks are troublesome since they exist in a region of the spectrum which is potentially very useful for optical communications: there are good LED and laser sources in the region (GaAs) and it is also, as we shall see in the next section, a region of low dispersion, which means that its information-carrying capacity is very large.

Optical-fiber communications technology really began when, in the mid-1960s, it was realized [5] that the loss in silica "glass" was due to impurities which were removable by known processes: these impurities were mostly metallic

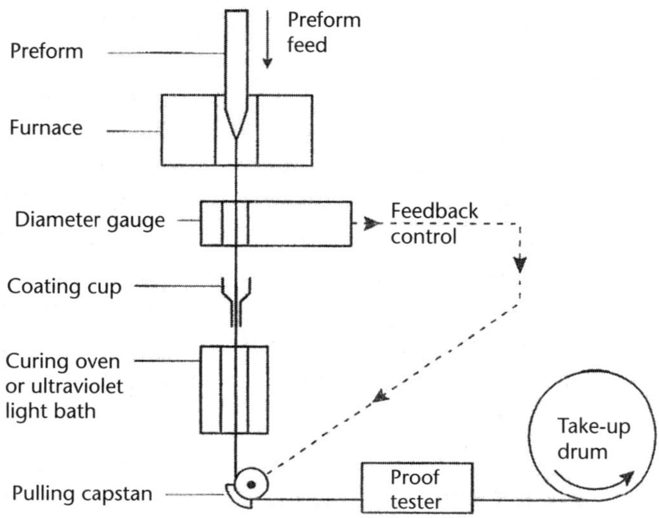

Figure 2.13 Schematic of a fiber-pulling rig.

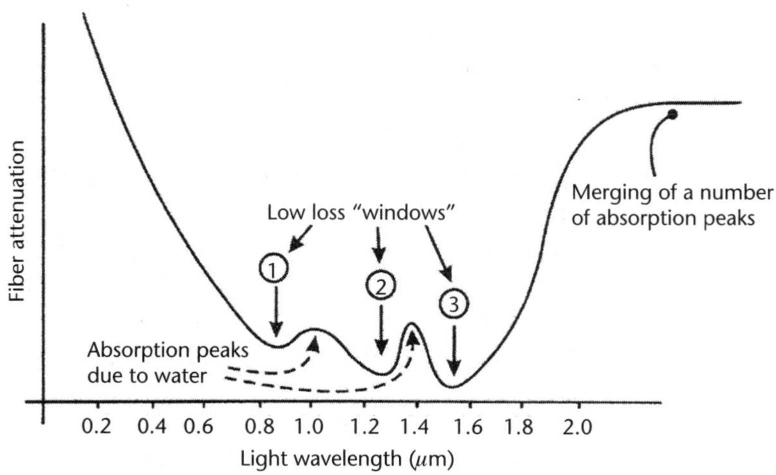

Figure 2.14 Absorption spectrum for silica fiber.

ions such as Fe^{3+}, Mn^{2+}, Ni^{2+}, Co^{2+}, and so on. Having removed these, the problem of the O-H resonance remained: this was the result of residual water in the structure, and it proved very difficult to remove. However, by the mid-1970s a concentrated attack on this material problem had given rise to silica of such purity that the secondary "water" peaks were hardly noticeable below 1.2 μm. The attenuation which remained was due almost entirely to Rayleigh scattering (~ $1/\lambda^4$), which is a fundamental property of the amorphous silica material structure and cannot be reduced substantially. Clearly, under these conditions, the larger the optical wavelength (the smaller the frequency), the smaller will be the attenuation, and the better will be the communications link. This remains true until we reach wavelengths in excess of ~ 1.55 μm when other resonances such as Si-O (fundamental material), Be-O and Ge-O (core dopant) start to give rise to absorption again. Thus 1.55 μm clearly is a good wavelength to use for communications. Losses as low as 0.2 dB km^{-1} can be achieved there. However, there are considerations other than just attenuation to consider in the choice of the working communications wavelength. One of these is the availability of suitable sources. Another problem is that of the dispersion. We shall now take a closer look at this last feature.

2.6.2 Optical-Fiber Dispersion

Optical material dispersion is a consequence of the variation of refractive index with optical frequency, and it has its origins in the same atomic absorption processes that give rise to the absorptive component of the attenuation spectrum.

Clearly, any optical energy propagating in a material medium will comprise a range of wavelengths. It is not possible to devise a source of radiation that has zero spectral width. Consequently, in the face of optical dispersion in the medium, different parts of the propagating energy will travel at different velocities; and if that energy is carrying information (i.e., it has been modulated in some way) that information will become distorted by the velocity differences. The further it travels, the greater will be the distortion; the greater the wavelength spread, the greater will be the distortion; the greater the dispersion power of the medium, again, the greater will be the distortion. For good communications we need, therefore, to choose our sources, wavelengths, and materials very carefully, and in order to make these choices we must understand the processes involved.

In optical fibers and in all other optical waveguides there are three types of dispersion: modal dispersion (in multimode guides only), material dispersion (which we already know something about), and waveguide dispersion (a consequence of the guide's geometry).

The effect of dispersion in a waveguide is to limit its communications-carrying capacity (i.e., its bandwidth). This is seen most readily by considering a digital communications system—that is, one which transmits information by means of a stream of pulses (Figure 2.15). (The presence of a pulse indicates a "1," the absence of one indicates a "0," and the stream thus comprises a digital coding of the information to be transmitted.) A stream of clear, distinct pulses is launched into the fiber (for example) by modulating a laser source. As the pulses propagate down the fiber, the spread of optical wavelengths of which

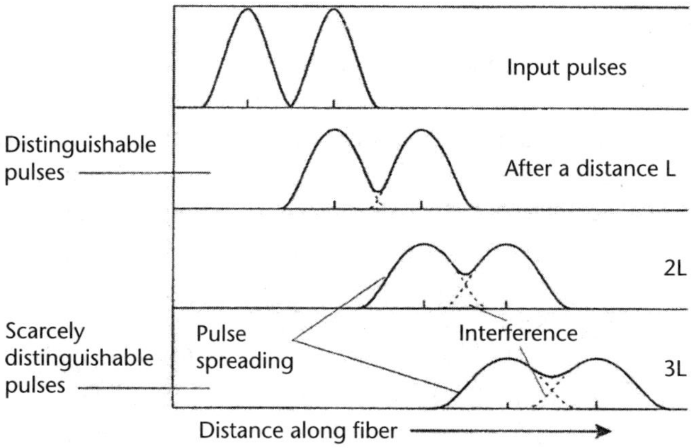

Figure 2.15 Effect of fiber dispersion on a pulsed optical signal.

they are comprised will be acted upon by the dispersive effects in the fiber, and the result will be a broadening of the pulses (Figure 2.15). When the broadening has become so great that it is no longer possible to distinguish between two successive pulses, the communications link fails.

Clearly, for a given dispersive power, the broadening will increase linearly with distance. Hence,

$$\frac{\Delta\tau}{L} = \text{constant}$$

where $\Delta\tau$ is the broadening (in time) of the pulse over a fiber length, L. Now the bit-rate, effectively the digital bandwidth, which the fiber length, L, can carry, will be $1/\Delta\tau$, since the spacing between pulses, $\Delta\tau$, will be closed up by the dispersion when the broadening is equal to $\Delta\tau$. Hence we have

$$BL = \text{constant} \tag{2.12}$$

where B is the allowable bit rate in pulses (or bits, i.e., binary digits) per second. It gives a good idea of the capacity of modern optical communications systems to know that this bit rate is usually quoted in megabits/second ($\text{Mb} \cdot \text{s}^{-1}$) or Gigabits/second ($\text{Gb} \cdot \text{s}^{-1}$).

The result of dispersion is thus to impose a bandwidth × distance limitation: the greater the distance, the smaller is the bandwidth that can be transmitted for a given fiber, and vice versa.

Let us look now at the particular causes of dispersion in optical fibers.

2.6.2.1 Modal Dispersion

Modal dispersion was introduced briefly in Section 2.5. This dispersion exists only in multimode fibers, since it results from the differing velocities of the range of modes supported by the fiber. This is not a material dispersion but results from the fiber's structure. Optical energy is launched into the fiber and will be launched into many, perhaps all, of the modes supported by the fiber. The effect of modal dispersion clearly will depend on how the propagating energy is distributed among the possible modes, and this will vary along the fiber as the energy redistributes itself according to local conditions (e.g., bends, joints, and so on). In order to get a feel for its order of magnitude, however, we can very easily calculate the difference in time of flight, over a given distance, between the fastest and slowest modes supported by the fiber.

The fastest mode will be that which travels (almost) straight down the fiber, along the axis (Figure 2.16). This will have the velocity of the unbounded core medium, c_0/n_1.

The slowest mode will be that which is represented by a ray which is incident on the core/cladding boundary at the TIR angle (for any greater angle the ray will not be guided). Clearly (Figure 2.16) this ray travels at velocity $(c_0/n_1)\sin\vartheta_c$, where ϑ_c is the critical angle.

Since we have

$$\sin\vartheta_c = \frac{n_2}{n_1}$$

it is easily seen that the two times of flight along a distance L of fiber are

$$\tau_f = L\frac{n_1}{c_0};\ \tau_s = \frac{Ln_1}{c_0\sin\vartheta_c} = \frac{Ln_1^2}{n_2}$$

Hence,

$$\Delta\tau = \tau_s - \tau_f = \frac{L}{c_0}\frac{n_1}{n_2}(n_1 - n_2)$$

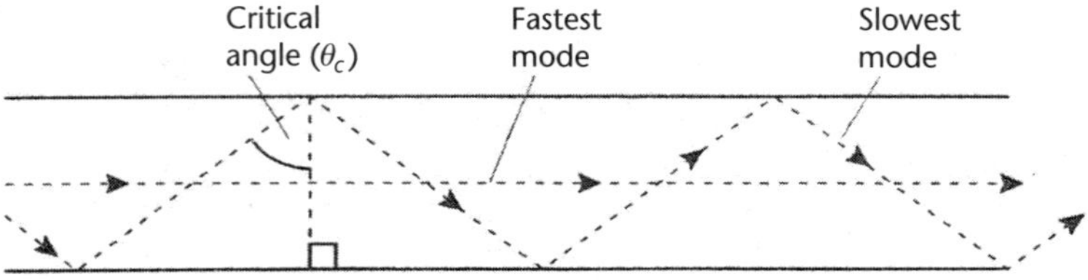

Figure 2.16 Modal dispersion.

And since

$$n_1 \approx n_2 \left(\frac{n_1 - n_2}{n_1} \sim 0.01 \right)$$

then

$$\Delta \tau \approx \frac{L}{c_0} \Delta n \tag{2.13}$$

where Δn is the difference in refractive index between core and cladding. Equation (2.13) is a clear specific example of the general (2.12), for we have, from (2.13),

$$\frac{\Delta \tau}{L} = \frac{\Delta n}{c_0}; \; B = \frac{1}{\Delta \tau}$$

Hence,

$$BL = \frac{c_0}{\Delta n}$$

The RHS is thus a constant for a given fiber. Typically, with a refractive index difference of ~ 0.01 we have

$$BL = 3 \times 10^{10} \text{ Hz m} = 30 \text{ MHz} \cdot \text{km}$$

Hence, for such a fiber, only a 30-MHz bandwidth is available over a 1-km length; only 3 MHz over 10 km, and so on. Multimode fiber clearly is seriously limited in its bandwidth capability.

It is instructive also to relate B to the amount of optical power which can be launched into the fiber from a given source. From Section 2.5 we know that the numerical aperture (NA) is given by

$$\text{NA} = \left(n_1^2 - n_2^2\right)^{1/2} = [(n_1 - n_2)(n_1 + n_2)]^{1/2}$$

or

$$(\text{NA})^2 \approx (\Delta n)(2n_1)$$

since

$$\frac{(n_1 - n_2)}{n_1} \sim 0.01$$

Substituting for Δn in (2.13) we find

$$\Delta\tau = \frac{L}{2n_1 c_0}(\mathrm{NA})^2$$

Now the numerical aperture is a measure of the ease with which the fiber will accept light from a source. We can see from Figure 2.17 that the solid angle of acceptance is just $\pi(\mathrm{NA})^2$. Hence the greater the value of $(\mathrm{NA})^2$, the greater will be the launched power. If we also assume that the noise on the received signal is independent of fiber length (a fair assumption since almost all the noise will be shot and thermal noise generated in the receiver), then it follows that the detection signal-to-noise ratio (SNR) is proportional to the launched power (for a given fiber length) and thus to $(\mathrm{NA})^2$. Hence

$$\mathrm{SNR} \sim (\mathrm{NA})^2$$

$$\frac{1}{\Delta\tau} \sim B \sim \frac{2n_1 c_0}{L(\mathrm{NA})^2}$$

and

$$SNR \times B \sim 2n_1 c_0 / L$$

Thus, for a given fiber length, the product of SNR and bandwidth is also a constant. Increasing the NA (for example) may increase the power into the fiber, but this is at the expense of a reduced bandwidth, owing to the increased modal dispersion which results from the greater NA. Such relationships are generally true in communications systems but are especially easy to appreciate

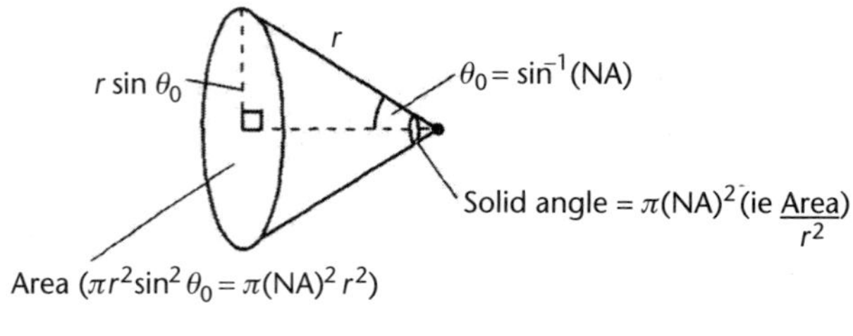

Figure 2.17 Solid angle for fiber's light acceptance.

for multimode optical-fiber links. These relationships allow a glimpse of the kinds of compromise that must be faced by optical-fiber communications systems designers. In order to minimize multimode dispersion, and thus maximize the bandwidth for a given fiber length, it is clear that the number of modes must be minimized. The absolute minimum number that we can have is one: a monomode fiber. It is for this reason that monomode (or single mode) fibers are the preferred medium for optical-fiber communications: only quite short distance (<1 km) links now employ multimode fibers.

However, in monomode fibers other sources of dispersion become important. These are overwhelmed by modal dispersion in multimode fibers, but when this is removed, it is these other dispersion effects that limit the bandwidth of the communications system. These other effects will now be considered.

2.6.2.2 Material Dispersion

Material dispersion is due to the fact that the refractive index of any optical material will vary with wavelength, owing to the structure of the atomic resonances.

From Section 1.7 we know that a variation of refractive index with wavelength will give rise to a group velocity, different from the phase velocity, given by

$$c_g = \frac{d\omega}{dk}$$

which, with the aid of the relation

$$\omega = \frac{c_0}{n} k$$

where n is the refractive index of the material, could be translated into

$$c_g = \frac{c_0}{n}\left(1 - \frac{k}{n}\frac{dn}{dk}\right)$$

where c_g is the velocity with which the optical energy travels.

Clearly, c_g will not be a constant across the source spectrum unless dn/dk is constant across it (i.e., unless n is a linear function of wavelength). If it is not, then different portions of the source spectrum travel at different velocities and this will result in, among other effects, the broadening of an optical pulse.

The time taken for energy to travel a distance L when its group velocity is c_g is

$$\tau = \frac{L}{c_g}$$

Hence the spread of times for a source spectrum of width $\delta\omega$ will be

$$\Delta\tau = L\frac{d\left(c_g^{-1}\right)}{d\omega}\,\delta\omega = L\frac{d^2k}{d\omega^2}\,\delta\omega \tag{2.14}$$

Since $k = 2\pi/\lambda = n\omega/c_0$, this can also be expressed as

$$\Delta\tau = \frac{L}{c_0}\lambda\left(\frac{d^2n}{d\lambda^2}\right)_\lambda \delta\lambda$$

where $(d^2n/d\lambda^2)_\lambda$ is the value of $(d^2n/d\lambda^2)$ at the wavelength λ. Again it can be noted that, with $B \sim 1/\Delta\tau$

$$BL = \frac{c_0}{\lambda\left(\dfrac{d^2n}{d\lambda^2}\right)\delta\lambda}$$

and is a constant for a given source and material.

It is clear that, for maximum bandwidth × distance product, $d^2n/d\lambda^2$, λ, and $\delta\lambda$ must be as small as possible. Now, clearly (λ/c_0) $(d^2n/d\lambda^2)$ (at the wavelength λ) is a characteristic of the material. Its variation with λ is shown in Figure 2.18 for silica.

From this it can be seen that $d^2n/d\lambda^2 = 0$ at $\lambda \sim 1.28\ \mu$m.

It follows that this is the preferred wavelength for maximum bandwidth in a silica fiber. It is extremely fortuitous that this wavelength also corresponds to a minimum in the absorption spectrum for silica (see Figure 2.14), thus also giving low attenuation at this wavelength. It was the combination of these two factors that led to the rapid progress of monomode optical-fiber communications technology in the 1980s.

In order to take maximum advantage of the bandwidth capability at this wavelength it is necessary, of course, also to minimize $\delta\lambda$, the optical source's spectral width. This requirement has led to the development of relatively higher power (~ 100 mW) narrowband (~ 0.1 nm) semiconductor laser sources and, more recently, optical-fiber lasers.

Of course, $d^2n/d\lambda^2$ will not be zero over the whole width of the source. The best we can do is ensure that it is zero at or around the center of the spectrum, λ_0.

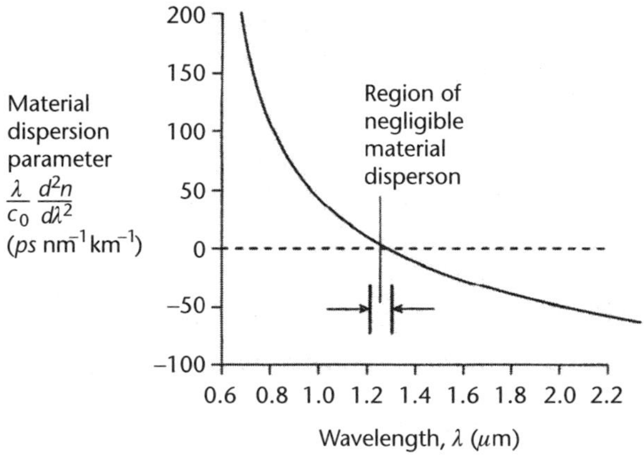

Figure 2.18 The material dispersion "zero" for silica.

Let us insert some typical numbers into our equations to get a feel for practicalities.

Let us assume that the light source is a semiconductor laser with a wavelength spread of ~ 1 nm. From Figure 2.18 we can see that the mean material dispersion for a range of 1 nm around the zero point (1.28 μm) is ~ 3 ps nm^{-1} · km^{-1}. Hence, $\Delta\tau \approx 3 \times 10^{-12}$ km^{-1} or $BL \approx 330$ Gbs^{-1} · km.

Therefore, for a link of length 100 km, for example, we shall have a bandwidth of 3.3 Gbs^{-1}. This is a respectable bandwidth but it should be possible to do better over this distance. The most obvious approach for improvement is to reduce still further the spectral width of the source, to ~ 0.1 nm, say, where 33 Gbs^{-1} will be available over the same distance. Before settling for this, however, there is yet another source of dispersion to worry about in regard to monomode fibers. This is known as waveguide dispersion and will now be considered.

2.6.2.3 Waveguide Dispersion

When considering planar waveguides in Section 2.2 and cylindrical waveguides in Section 2.4, the relationships between ω and β (Figures 2.3 and 2.8) were seen to be nonlinear for any given mode and, in particular, for the lowest order mode. This lowest order mode is, of course, the only mode propagating in a monomode fiber. The nonlinearity derives from the necessity to satisfy the guide's boundary condition at the different optical frequencies, and especially from the fact that the angle at which rays strike the boundary varies with frequency, which, in turn, leads to a different phase change on TIR; this latter is a complex function of the angle of incidence (Section 1.4).

From the discussion in the last section on pulse spreading, it was noted in (2.14) that the spread of arrival times, after a distance L, for a pulse whose energy is spread over a spectral width $\delta\omega$, is

$$\Delta\tau = L\frac{d^2k}{d\omega^2}\delta\omega$$

Clearly, for the present case, the guide wavenumber β must be substituted for k to give

$$\Delta\tau = L\frac{d^2\beta}{d\omega^2}\delta\omega$$

Hence, in the case of waveguide dispersion, $\Delta\tau$ can be calculated from the β/ω curves (the dispersion curves) for any given mode and, in particular, for the lowest mode in the monomode case.

The physical origins of this form of dispersion lie in two factors. First, as the frequency varies for a given mode, the change of angle of incidence at the boundary means that the ray progresses down the guide at a different velocity (the velocity varies as sin ϑ, in fact). Second, as the frequency varies and the angle at the boundary varies, the penetration of the evanescent wave into the second medium (cladding) varies, in accordance with (2.5). Hence the relative amounts of power in the guiding channel (e.g., core) and in the outer medium's evanescent field will vary with frequency. Since the refractive indices for the two media are different, it is unsurprising that this leads to a variation of the guided wave's effective refractive index with frequency.

It transpires, in the case of cylindrical waveguides [3] that the ratio of the power in the core to that in the cladding for the lowest order mode is given, as a first approximation, by

$$\frac{P_{\text{core}}}{P_{\text{cladding}}} = V^2 \tag{2.15}$$

where

$$V = \frac{2\pi a}{\lambda}\left(n_1^2 - n_2^2\right)^{1/2}$$

Hence, as λ rises, the power in the core increasingly transfers to the cladding. This, again, can readily be justified qualitatively from the physics: as

λ rises, the angle of incidence (ϑ) decreases (i.e., the ray becomes more steeply inclined to the axis), as required by (2.3a). As ϑ decreases, the penetration of the evanescent wave increases (2.10b). Hence the cladding medium has greater effect at higher wavelengths (lower β) and the guide refractive index tends towards that of the cladding, n_2 [Figure 2.19(a)]. Clearly, all these arguments are reversed as the wavelength decreases. In particular, a decreasing wavelength causes the refractive index to tend towards that of the core; indeed, for very small wavelengths—very much less than the core diameter—the wave is, to first order, unaware of the boundary between the media and propagates as if it were doing so in an unrestricted core medium, of refractive index n_1.

This waveguide dispersion can, in fact, be very useful, for it can be arranged to oppose the material dispersion. The waveguide dispersion depends upon the fiber geometry (2.15), so, by choosing the geometry appropriately, the dispersion minimum for the fiber can be shifted from its value defined by material dispersion to another convenient (but quite close) wavelength [Figure 2.19(b)]. Fiber that has been adjusted in this way is called *dispersion-shifted fiber*, and this stratagem

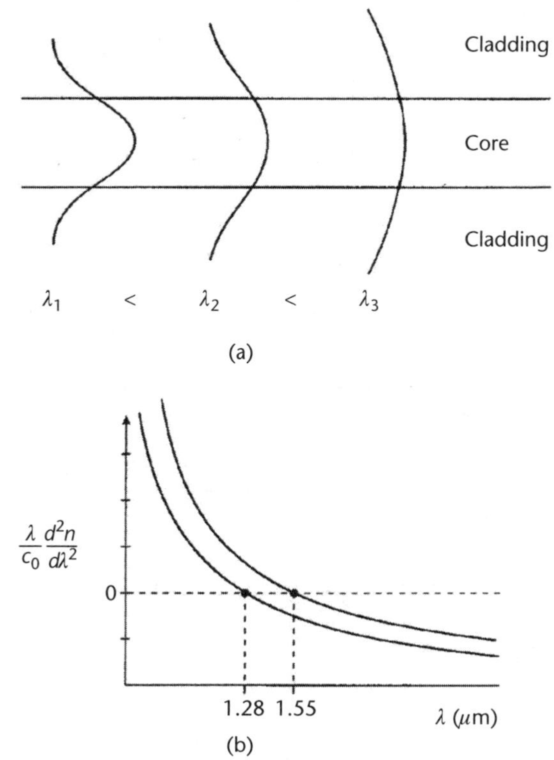

Figure 2.19 Waveguide dispersion: (a) variation with wavelength of power distribution in a fiber; and (b) dispersion shifting for silica fiber.

is used to move the wavelength of the dispersion minimum from its unrestricted value of 1.28 μm to 1.55 μm, for example, where the attenuation due to absorption is lower. Unrepeatered trunk telecommunications systems of several hundred kilometer lengths can be installed using such fiber.

2.7 Conclusions

Optical waveguiding is of primary importance to the optoelectronic designer. With its aid, it is possible to confine light and to direct it to where it is needed, over short, medium, and long distances.

Furthermore, with the advantage of confinement, it is possible to control the interaction of light with other influences, such as electric, magnetic, or acoustic fields, which may be needed to impress information upon it. Control also can be exerted over its intensity distribution, its polarization state and its nonlinear behavior.

In short, optical waveguiding is crucial to the control of light. For the designers of devices and systems this control is essential.

References

[1] Syms, R., and J. Cozens, *Optical Guided Waves and Devices*, New York: McGraw-Hill, 1992, Chapter 6.

[2] Kaplan, W., *Advanced Mathematics for Engineers*, Reading, MA: Addison-Wesley, 1981, Chapter 12.

[3] Adams, M. J., *An Introduction to Optical Waveguides*, New York: John Wiley, 1981, Chapter 7.

[4] Senior, J. M., *Optical Fiber Communications: Principles and Practice*, 2nd ed., Englewood Cliffs, NJ: Prentice-Hall, 1992, Chapter 4.

[5] Kao, K. C., and G. A. Hockham, "Dielectric Fiber Surface Waveguides for Optical Frequencies," *Proc. IEE*, Vol. 113, No. 7, 1966, pp. 1151–1158.

Selected Bibliography

Marz, R., *Integrated Optics: Design and Modeling*, Norwood, MA: Artech House, 1995.

Midwinter, J. E., *Optical Fibers for Transmission*, New York: John Wiley, 1978.

Najafi, S. I., *Introduction to Glass Integrated Optics*, Norwood, MA: Artech House, 1992.

3

Elements of Polarization Optics

3.1 Introduction

In order to understand the polarization behavior of optical fibers, it is first necessary to examine the general subject of polarization optics in more detail. In Section 1.2.3 the basic ideas of optical polarization were introduced. In the present chapter we shall expand and develop the ideas in regard to their general position in optical physics, before going on, in the following chapters, to look at the particular relevance and application of the ideas to optical fibers.

We know that the electric and magnetic fields, for a freely propagating light wave, lie transversely to the propagation direction and orthogonally to each other. Normally, when discussing polarization phenomena, we fix our attention on the electric field, since it is this which has the most direct effect when the wave interacts with matter.

In saying that an optical wave is polarized we are implying that the direction of the optical field is either constant or is changing in an ordered, prescribable manner. In general, the tip of the electric vector circumscribes an ellipse, performing a complete circuit in a time equal to the period of the wave, or in a distance of one wavelength. Clearly, the two descriptions are equivalent in this respect.

It is also clear that the optical field can only change in any given, ordered fashion as long as the wave remains coherent, for the coherence determines the length, or time, over which the phase relationships, between orthogonal field components, remain constant. This implies that, for any other than the linear polarization state, any given polarization state can only be retained for the coherence length, or coherence time, of the light.

As is well known, linearly polarized light can conveniently be produced by passing any light beam through a sheet of polarizing film. This is a material that transmits light of one linear polarization (the *acceptance* direction) to a much greater extent (~ 1,000 times) than the orthogonal linear polarization, thus, effectively, allowing just one linear polarization state to pass. The material's properties result from the fact that it consists of long-chain polymeric molecules aligned in one direction (the acceptance direction) by stretching a plastic, and then stabilizing it. Electrons can move more easily along the chains than transversely to them, and thus the optical wave transmits easily only when its electric field lies along this acceptance direction. The material is cheap and allows the use of large optical apertures. It thus provides a convenient means whereby, for example, a specific linear polarization state can be defined; this state then provides a ready polarization reference that can be used as a starting point for other manipulations. In order to study these manipulations and other aspects of polarization optics, we shall begin by looking more closely at the polarization ellipse.

3.2 The Polarization Ellipse

In Section 1.2.3 the most general form of polarized light wave propagating in the *Oz* direction was derived from the two linearly polarized components in the *Ox* and *Oy* directions (Figure 3.1):

$$E_x = e_x \cos(\omega t - kz + \delta_x) \tag{3.1a}$$

$$E_y = e_y \cos(\omega t - kz + \delta_y)$$

If we eliminate $(\omega t - kz)$ from these equations we obtain the expression:

$$\frac{E_x^2}{e_x^2} + \frac{E_y^2}{e_y^2} - \frac{2E_xE_y}{e_xe_y}\cos(\delta_y - \delta_x) = \sin^2(\delta_y - \delta_x) \tag{3.1b}$$

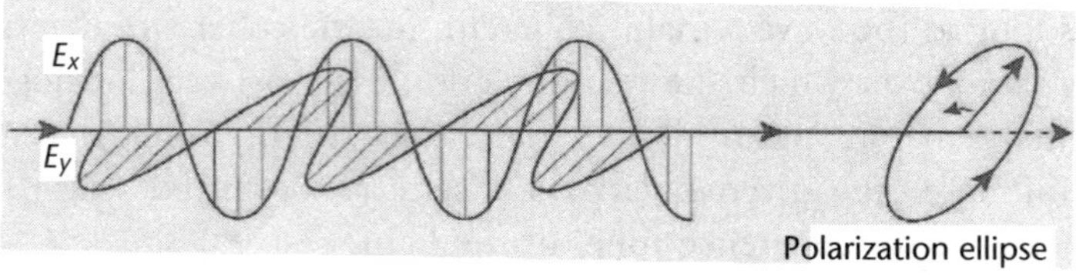

Figure 3.1 Components for an elliptically polarized wave.

which is the ellipse (in the variables E_x, E_y) circumscribed by the tip of the resultant electric vector at any one point in space over one period of the combined wave. This can only be true, however, as has already been indicated, if the phase difference $(\delta_y - \delta_x)$ is constant in time, or, at least, changes only slowly when compared with the speed of response of the detector. In other words, we say that the two waves must have a large mutual *coherence*. If this were not so, then relative phases and hence resultant field vectors would vary randomly within the detector response time, giving no ordered pattern to the behavior of the resultant field and thus presenting to the detector what would be, essentially, unpolarized light.

Assuming that the mutual coherence is good, we may investigate further the properties of the polarization ellipse.

Note, first, that the ellipse always lies in the rectangle shown in Figure 3.2, but that the axes of the ellipse are not parallel with the original x, y directions.

The ellipse is specified as follows: with e_x, e_y, δ $(= \delta_y - \delta_x)$ known, we then define $\tan \beta = e_y / e_x$.

The orientation of the ellipse, α, is given by (see Appendix C)

$$\tan 2\alpha = \tan 2\beta \cos \delta$$

Semi-major and semi-minor axes a, b are related by

$$e_x^2 + e_y^2 = a^2 + b^2 \sim I$$

The ellipticity of the ellipse, e, is given by

$$e = \tan \chi = \pm b/a$$

(the sign determines the sense of the rotation) where $\sin 2\chi = -\sin 2\beta \sin \delta$.

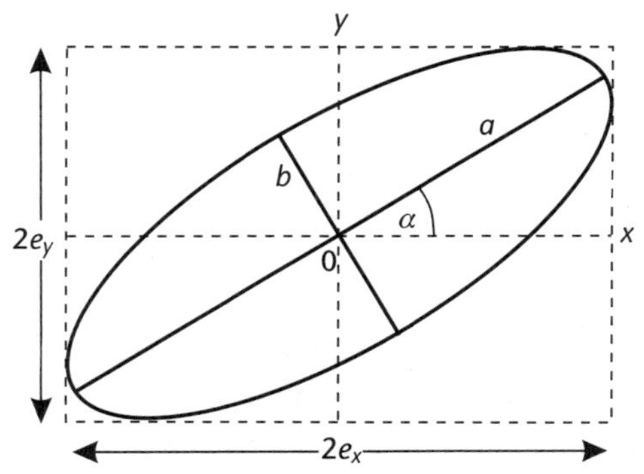

Figure 3.2 The polarization ellipse.

We should note also that the electric field components along the major and minor axes are always in quadrature (i.e., $\pi/2$ phase difference, the sign of the difference depending on the sense of the rotation).

Linear and circular states of polarization may be regarded as special cases where the polarization ellipse degenerates into a straight line or a circle, respectively.

A linear state is obtained with the components in (3.1a) when either

$$\left.\begin{matrix} e_x = 0 \\ e_y \neq 0 \end{matrix}\right\} \quad \text{linearly polarized in } Oy \text{ direction}$$

$$\left.\begin{matrix} e_x \neq 0 \\ e_y = 0 \end{matrix}\right\} \quad \text{linearly polarized in } Ox \text{ direction}$$

or

$$\delta_y - \delta_x = m\pi$$

where m is an integer. In this last case the direction of polarization will be at an angle,

$$+\tan^{-1}\left(\frac{e_y}{e_x}\right) \qquad m \text{ even}$$

$$-\tan^{-1}\left(\frac{e_y}{e_x}\right) \qquad m \text{ odd}$$

with respect to the Ox axis.

A circular state is obtained when

$$e_x = e_y$$

and

$$(\delta_y - \delta_x) = (2m + 1)\pi/2$$

that is, in this case the two component waves have equal amplitudes and are in phase quadrature. The waves will be right-hand circularly polarized when m is even and left-hand circularly polarized when m is odd.

Light can become polarized as a result of the intrinsic directional properties of matter: either the matter that is the original source of the light, or the matter through which the light passes. These intrinsic material directional properties are the result of directionality in the bonding that holds together the atoms of which the material is made. This directionality leads to variations in the response of the material according to the direction of an imposed force, be it electric, magnetic, or mechanical. The best known manifestation of directionality in solid materials is the crystal, with the large variety of crystallographic forms, some symmetrical, some asymmetrical. The characteristic shapes that we associate with certain crystals result from the fact that they tend to break preferentially along certain planes, known as cleavage planes, which are those planes between which atomic forces are weakest.

It is not surprising, then, to find that directionality in a crystalline material is also evident in the light that it produces, or is impressed upon the light that passes through it.

In order to understand the ways in which we may produce polarized light, control it and use it, we must make a gentle incursion into the subject of crystal optics.

3.3 Crystal Optics

Light propagates through a material by stimulating the elementary atomic dipoles to oscillate and thus to radiate. In our previous discussions the forced oscillation was assumed to take place in the direction of the driving electric field, but in the case of a medium whose physical properties vary with direction, an anisotropic medium, this is not necessarily the case. If an electron in an atom or molecule can move more easily in one direction than another, then an electric field at some arbitrary angle to the preferred direction will move the electron in a direction which is not parallel with the field direction (Figure 3.3). As a result, the direction in which the oscillating dipole's radiation is maximized (i.e., normal to its oscillation direction) is not the same as that of the driving wave.

The consequences, for the optics of anisotropic media, of this simple piece of physics are complex.

Immediately we can see that the already-discussed relationship between the electric displacement **D** and the electric field **E**, for an *isotropic* (i.e., with no directionality) medium,

$$\mathbf{D} = \epsilon_R \epsilon_o \mathbf{E}$$

must be more complex for an anisotropic medium; in fact, the relation must now be written in the form (for any, arbitrary three orthogonal directions *Ox*, *Oy*, *Oz*):

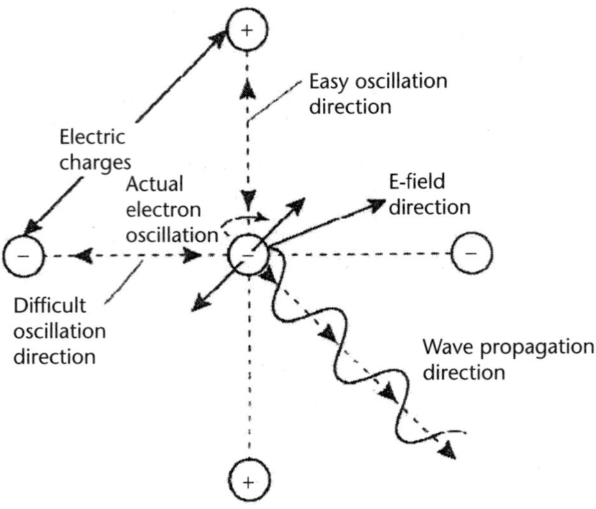

Figure 3.3 Electron response to electric field in an anisotropic medium.

$$D_x = \epsilon_o(\epsilon_{xx}E_x + \epsilon_{xy}E_y + \epsilon_{xz}E_z)$$
$$D_y = \epsilon_o(\epsilon_{yx}E_x + \epsilon_{yy}E_y + \epsilon_{yz}E_z)$$
$$D_z = \epsilon_o(\epsilon_{zx}E_x + \epsilon_{zy}E_y + \epsilon_{zz}E_z)$$

Clearly, what is depicted here is an array which describes the various electric field susceptibilities in the various directions within the crystal: ϵ_{ij} (a scalar quantity) is a measure of the effect which an electric field in direction $\boldsymbol{j}$ has in direction $\boldsymbol{i}$ within the crystal, that is, the ease with which it can move electrons in that direction and thus create a dipole moment.

The array can be written in the abbreviated form

$$D_i = \epsilon_o \epsilon_{ij} E_j \qquad (i, j = x, y, z)$$

and ϵ_{ij} is now a *tensor* known, in this case, as the permittivity tensor. A tensor is a physical quantity which characterizes a particular physical property of an anisotropic medium, and takes the form of a matrix. Clearly **D** is not now (in general) parallel with **E**, and the angle between the two also will depend upon the direction of **E** in the material.

Now it can be shown from energy considerations that the permittivity tensor is symmetrical, that is, $\epsilon_{ij} = \epsilon_{ji}$. Also, symmetrical tensors can be cast into their diagonal form by referring them to a special set of axes (the principal axes) which are determined by the crystal structure [1]. When this is done, we have

$$\begin{pmatrix} D_x \\ D_y \\ D_z \end{pmatrix} = \epsilon_0 \begin{pmatrix} \epsilon_{xx} & 0 & 0 \\ 0 & \epsilon_{yy} & 0 \\ 0 & 0 & \epsilon_{zz} \end{pmatrix} \begin{pmatrix} E_x \\ E_y \\ E_z \end{pmatrix}$$

The new set of axes, *Ox*, *Oy*, *Oz*, is now this special set.

Suppose now that $\mathbf{E} = E_x \mathbf{i}$; that is, we have, entering the crystal, an optical wave whose **E** field lies in one of these special crystal directions.

In this case we simply have

$$D_x = \epsilon_o \epsilon_{xx} E_x$$

as our tensor relation, and ϵ_{xx} is, of course, a scalar quantity. In other words, we have **D** parallel with **E**, just as for an isotropic material, and the light will propagate, with refractive index $\epsilon_{xx}^{1/2}$, perfectly normally. Furthermore, the same will be true for

$$\mathbf{E} = E_y \mathbf{j} \left(\text{refractive index } \epsilon_{yy}^{1/2}\right)$$

$$\mathbf{E} = E_z \mathbf{k} \left(\text{refractive index } \epsilon_{zz}^{1/2}\right)$$

Before going further we should note an important consequence of all this: the refractive index varies with the direction of **E**. If we have a wave traveling in direction *Oz*, its velocity now will depend upon its polarization state: if the wave is linearly polarized in the *Ox* direction it will travel with velocity $c_0 / \epsilon_{xx}^{1/2}$, while if it is linearly polarized in the *Oy* direction its velocity will be $c_0 / \epsilon_{yy}^{1/2}$. Hence the medium is offering two refractive indices to the wave traveling in this direction: we have the phenomenon known as double refraction or *birefringence*. A wave which is linearly polarized in a direction at 45° to *Ox* will split into two components, linearly polarized in directions *Ox* and *Oy*, the two components traveling at different velocities. Hence the phase difference between the two components will steadily increase and the composite polarization state of the wave will vary progressively from linear to circular and back to linear again.

This behavior is, of course, a direct consequence of the basic physics which was discussed earlier: it is easier, in the anisotropic crystal, for the electric field to move the atomic electrons in one direction than in another. Hence, for the direction of easy movement, the light polarized in this direction can travel faster than when it is polarized in the direction for which the movement is more sluggish. Birefringence is a long word, but the physical principles that underlie it really are very simple. It follows from these discussions that an anisotropic medium may be characterized by means of three refractive indices, corresponding to polarization directions along Ox, Oy, Oz, and that these will have values $\epsilon_{xx}^{1/2}$, $\epsilon_{yy}^{1/2}$, $\epsilon_{zz}^{1/2}$, respectively. We can use this information to determine the refractive index (and thus the velocity) for a wave in any direction with any given linear polarization state.

To do this we construct an *index ellipsoid*, or *indicatrix* as it is sometimes called (see Figure 3.4), from the form of the permittivity tensor for any given crystal. This ellipsoid has the following important properties.

Suppose that we wish to investigate the propagation of light, at an arbitrary angle to the crystal axes (polarization as yet unspecified). We draw a line, OP, corresponding to this direction within the index ellipsoid, passing through its center O (Figure 3.4). Now we construct the plane, also passing through O, which lies at right angles to the line. This plane will cut the ellipsoid in an ellipse. This ellipse has the property that the directions of its major and minor axes define the directions of linear polarization for which **D** and **E** are parallel for this propagation direction, and the lengths of these axes OA and OB are equal to the refractive indices for these polarizations. Since these two linear polarization states are the only ones which propagate without change of polarization form for this crystal direction, they are sometimes referred to as the *eigenstates* or *polarization eigenmodes* for this direction, conforming to the matrix terminology of eigenvectors and eigenvalues.

The propagation direction we first considered, along Oz, corresponds, of course, to one of the axes of the ellipsoid, and the two refractive indices $\epsilon_{xx}^{1/2}$ and $\epsilon_{yy}^{1/2}$ are the lengths of the other two axes in the central plane normal to Oz.

The refractive indices $\epsilon_{xx}^{1/2}$, $\epsilon_{yy}^{1/2}$, $\epsilon_{zz}^{1/2}$, are referred to as the *principal* refractive indices, and we shall henceforth denote them n_x, n_y, n_z.

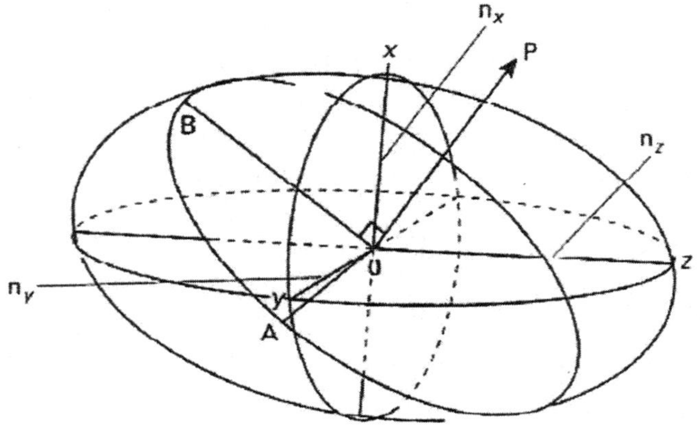

Figure 3.4 The index ellipsoid. OA and OB represent the linearly polarized eigenstates for the direction OP.

Several other points are worth noting. Suppose, first, that

$$n_x > n_y < n_z$$

It follows that there will be a plane which contains *Oz* for which the two axes of interception with the ellipsoid are equal (Figure 3.5). This plane will be at some angle to the *yz* plane and will thus intersect the ellipsoid in a circle. This means, of course, that, for the light propagation direction corresponding to the normal to this plane, all polarization directions have the same velocity; there is no double refraction for this direction. This direction is an *optic axis* of the crystal and there will, in general, be two such axes, since there must also be such a plane at an equal angle to the *yz* plane on the other side (see Figure 3.5). Such a crystal with two optic axes is said to be biaxial.

Suppose now that

$$n_x = n_y = n_o \text{ (say), the "ordinary" index}$$

and

$$n_z = n_e \text{ (say), the "extraordinary" index}$$

In this case one of the principal planes is a circle and it is the only circular section (containing the origin) that exists. Hence, in this case there is only one optic axis, along the *Oz* direction. Such crystals are said to be uniaxial (Figure 3.6). The crystal is said to be positive when $n_e > n_o$ and negative when

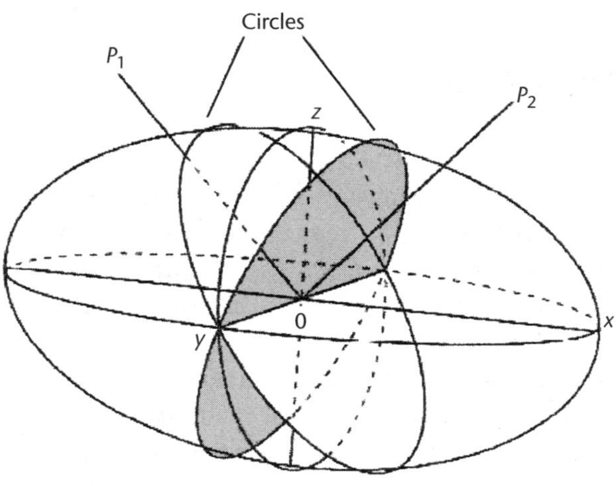

Figure 3.5 Ellipsoid for a biaxial crystal. P_1 and P_2 are the optical axes of the crystal.

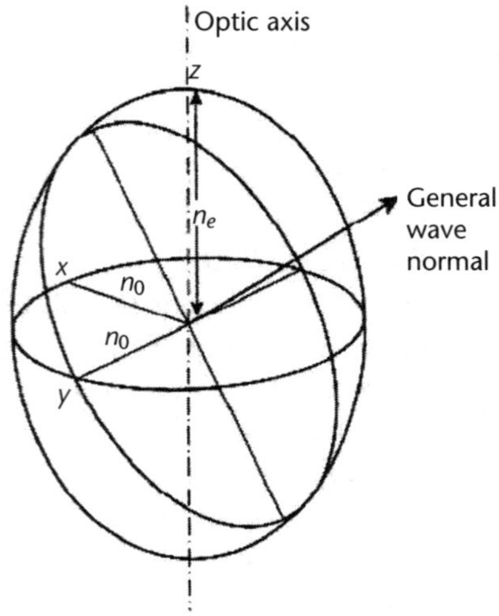

Figure 3.6 Ellipsoid for a (positive) uniaxial crystal.

$n_e < n_0$. For example, quartz is a positive uniaxial crystal, and calcite a negative uniaxial crystal. These features are, of course, determined by the crystal class to which these materials belong.

It is clear that the index ellipsoid is a very useful device for determining the polarization behavior of anisotropic media.

Let us now consider some practical consequences of all of the above.

3.4 Retarding Waveplates

Consider a positive uniaxial crystal plate (e.g., quartz) cut in such a way (Figure 3.7) as to set the optic axis parallel with one of the faces. Suppose a wave is incident normally on to this face. If the wave is linearly polarized with its **E** field parallel with the optic axis, it will travel with refractive index n_e as we have described; if it has the orthogonal polarization, normal to the optic axis, it will travel with refractive index n_0.

The two waves travel in the same direction through the crystal but with different velocities. For a positive uniaxial crystal $n_e > n_0$, and thus the light linearly polarized parallel with the optic axis will be a "slow" wave, while the one at right angles to the axis will be "fast." For this reason the two crystal directions are often referred to as the slow and fast axes.

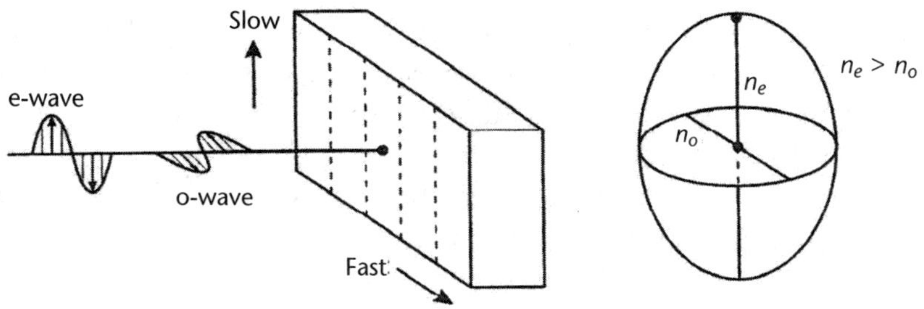

Figure 3.7 Plate with face parallel with the optic axis in quartz.

Suppose that the wave is linearly polarized at 45° to the optic axis. The phase difference between the components parallel with and orthogonal to the optic axis will now increase with distance l into the crystal according to

$$\varphi = \frac{2\pi}{\lambda}(n_e - n_o)l$$

Hence, if, for a given wavelength λ

$$l = \frac{\lambda}{4(n_e - n_o)}$$

Then,

$$\varphi = \frac{\pi}{2}$$

and the light emerges from the plate circularly polarized. We have inserted a phase difference of $\pi/2$ between the components, equivalent to a distance shift of $\lambda/4$, and the crystal plate, when of this thickness, is called a *quarter-wave* plate. It will (for an input polarization direction at 45° to the axes) convert linearly polarized light into circularly polarized light or vice versa. Clearly, a plate of the crystal also will act as a quarter-wave plate when, for any integer, N, we have

$$\varphi = (2N + 1)\pi/2$$

giving

$$l = (2N + 1) \cdot \lambda/4(n_e - n_o)$$

This implies a thicker plate, giving easier manufacturing and handling. The disadvantage of large N, however, is a greater temperature dependence of the thickness, and thus also of the quarter-wave condition. N is often referred to as the *order* of the plate.

If the input linear polarization direction lies at some arbitrary angle α to the optic axis, then the two components

$$E \cos \alpha$$

$$E \sin \alpha$$

will emerge with a phase difference of $\pi/2$. We noted in Section 3.2 that the electric field components along the two axes of a polarization ellipse were always in phase quadrature. It follows that these two components are now the major and minor axes of the elliptical polarization state that emerges from the plate. Thus the ellipticity of the ellipse (i.e., the ratio of the major and minor axis) is just tan α and by varying the input polarization direction α we have a means by which we can generate an ellipse of any ellipticity. The orientation of the ellipse will be defined by the direction of the optic axis of the waveplate [Figure 3.8(a)].

Suppose now that the crystal plate has twice the previous thickness and is used at the same wavelength. It becomes a *half-wave* plate. A phase difference of π is inserted between the components (linear eigenstates). The result of this is that an input wave which is linearly polarized at angle α to the optic axis will emerge still linearly polarized but with its direction now at $-\alpha$ to the axis.

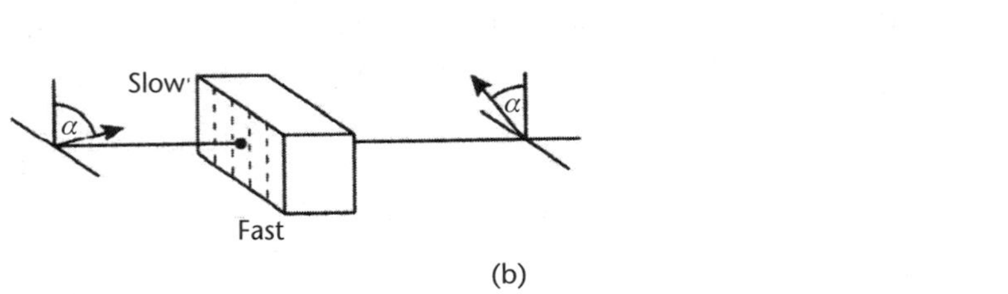

Figure 3.8 Polarization control with waveplates: (a) quarter-wave plate; and (b) half-wave plate.

The plate has rotated the polarization direction through an angle -2α. And, indeed, any input polarization ellipse will emerge with the same ellipticity but with its orientation rotated through -2α [Figure 3.8(b)].

It follows that, with the aid of these two simple plates, we can generate elliptical polarization of any prescribed ellipticity and orientation from linearly polarized light, which can itself be generated from any light source plus a simple polarizing sheet.

Equally valuable is the reverse process: that of the analysis of an arbitrary elliptical polarization state or its conversion to a linear state.

Suppose we have light of unknown elliptical polarization. By inserting a polarizing sheet and rotating it around the axis parallel to the propagation direction (Figure 3.9) we shall find a position of maximum transmission and an orthogonal position of minimum transmission. These are the major and minor axes of the ellipse (respectively) and the ratio of the two intensities at these positions will give the square of the ellipticity of the ellipse; that is,

$$e = \frac{b}{a} = \frac{E_b}{E_a} = \left(\frac{I_b}{I_a}\right)^{1/2}$$

Clearly, the orientation of the ellipse also is known since this is, by definition, just the direction of the major axis, and is given by the position at which the maximum occurs. In order to convert the elliptical state into a linear one, all we need is a quarter-wave plate (appropriate to the wavelength of the light used, of course). Since the components of the electric field along the major and minor axes of the ellipse are always in phase quadrature (see Section 3.2), the insertion of a quarter-wave plate with its axes aligned with the axes of the polarization ellipse will bring the components into phase or into antiphase, and the light

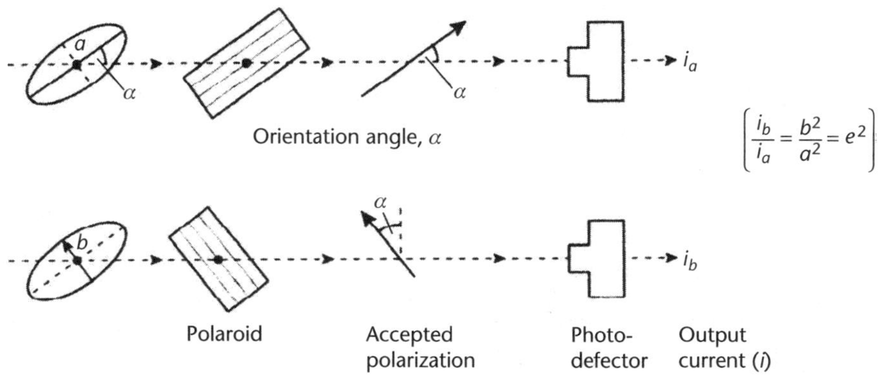

Figure 3.9 Determination of the polarization ellipse.

will thus become linearly polarized. The quarter-wave plate is used in conjunction with a following polaroid sheet (or prism polarizer) and the two are rotated (independently) about the propagation axis until the light is extinguished. The quarter-wave plate must then have the required orientation in line with the ellipse axes, since only when the light has become linearly polarized can the polarizer extinguish it completely. (If there are no positions for which the light is extinguished, then it is not fully polarized.)

Such are the quite powerful manipulations and analyses that can be performed with very simple devices. However, manual human intervention via rotation of plates is not always convenient or even possible.

In many cases polarization analysis and control must be done very quickly (perhaps in nanoseconds) and automatically, using electronic processing. For these cases more advanced polarization devices must be used and, in order to understand and use these, a more advanced theoretical framework is necessary. We shall introduce this in Section 3.10, and develop it further in Section 3.11.

3.5 A Variable Waveplate: The Soleil-Babinet Compensator

Consider the structure shown in Figure 3.10. A pair of wedges, cut from a crystal (e.g., quartz) so that its optic axis lies parallel with the front faces, rests on a rectangular block of the same crystal with its optic axis orthogonal to that of the wedges. The wedges may be moved laterally, as shown in the diagram, so that the total thickness of the upper slab, which the wedges comprise, is variable.

Consider now the incidence of a plane wave (1), normal to the upper surface and linearly polarized in a direction parallel with the optic axis of the wedges. Clearly it will travel through the wedges seeing a refractive index of n_e, and through the lower block with refractive index n_0. For a wave (2) with the orthogonal direction of polarization, the order of the refractive indices is reversed.

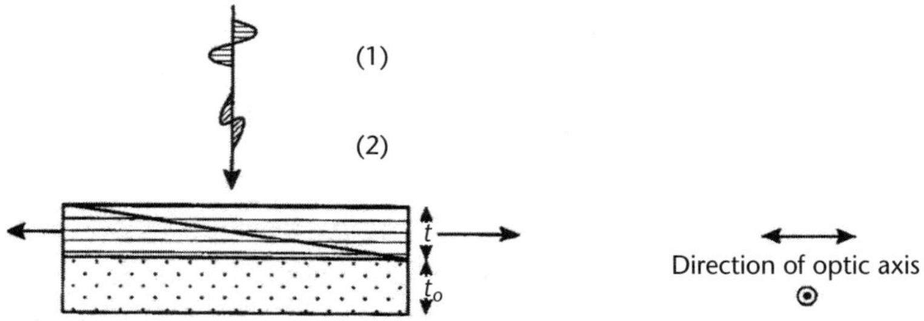

Figure 3.10 The Soleil-Babinet compensator.

Suppose that the "wedge" block thickness (variable) is t, while the lower block thickness (fixed) is t_0. Then it is clear that the phase delay suffered by the first wave will be

$$\varphi_1 = \frac{2\pi}{\lambda}(n_e t + n_o t_0)l$$

and by the second

$$\varphi_2 = \frac{2\pi}{\lambda}(n_o t + n_e t_0)l$$

giving a phase difference between the two

$$\varphi_2 - \varphi_1 = \frac{2\pi}{\lambda}(n_e - n_o)(t_0 - t)$$

This phase difference will be constant across that part of the aperture of the device which includes both wedges, and will be continuously variable from 0 to 2π, for any given wavelength, by sliding the wedges apart, and thus varying t. The device is known as a Soleil-Babinet (pronounced "Sollay-Babbinay") compensator (sometimes, but less commonly, a Babinet-Soleil compensator) and is very useful in both the control and analysis of optical polarization states. Clearly, the Soleil-Babinet compensator can be adjusted to form either a quarter-wave plate or a half-wave plate for any optical wavelength, if that is what is desired.

3.6 Polarizing Prisms

The same ideas as those just described are also useful in devices which produce linearly polarized light with a higher degree of polarization than a polarizing sheet is capable of, and without its intrinsic loss (even for the acceptance direction there is a significant loss). We shall look at just two of these devices, in order to illustrate the application of the ideas, but there are several others (these are described in most standard optics texts).

The first device is the Nicol prism, illustrated in Figure 3.11. Two wedges of calcite crystal are cut as shown, with their optic axes in the same direction (in the plane of the page) and cemented together with Canada balsam, a material whose refractive index at visible wavelengths lies midway between n_e and n_o. When unpolarized light enters parallel to the axis of the prim (as shown) and

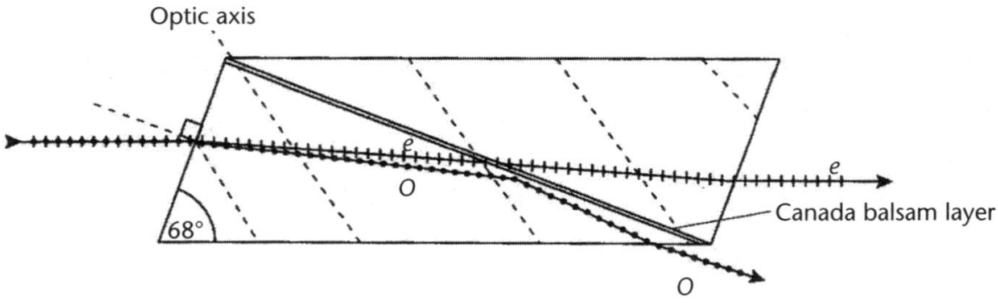

Figure 3.11 Action of the Nicol prism.

at an angle to the front face, it will split, as always, into the *e* and *o* components, each with its own refractive index, and thus each with its own refractive angle according to Snell's law. (Calcite is a negative uniaxial crystal so $n_o > n_e$.) When the light reaches the Canada balsam interface between the two wedges, it finds that the geometry and refractive indices have been arranged such that the ordinary (*o*) ray, with the larger deflection angle, strikes this interface at an angle greater than the total internal reflection (TIR) angle and is thus not passed into the second wedge, whereas the extraordinary (*e*) ray is so passed. Hence only the *e* ray emerges from the prism and this is linearly polarized. Thus we have an effective prism polarizer, albeit one of limited angular acceptance (~ 14°) since the TIR condition is quite critical in respect of angle of incidence.

The second prism we shall discuss is widely used in practical polarization optics: it is called the Wollaston prism, and is shown in Figure 3.12.

Again we have two wedges of positive (say) uniaxial crystal. They are equal in size, placed together to form a rectangular block (sometimes a cube), and have their optic axes orthogonal, as shown.

Consider a wave entering normally from the left. The *e* and *o* waves travel with differing velocities and strike the boundary between the wedges at the same angle. On striking the boundary, one of the waves sees a positive change

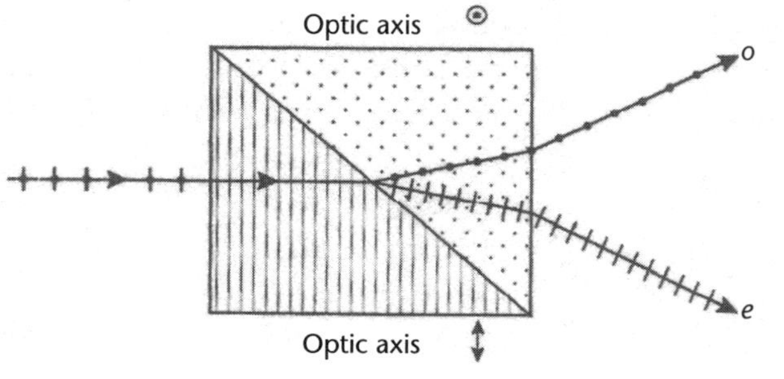

Figure 3.12 Action of the Wollaston prism.

in refractive index ($n_e - n_o$), the other a negative change ($n_o - n_e$), so that they are deflected, respectively, up and down (Figure 3.12) through equal angles. The *e* and *o* rays thus diverge as they emerge from the prism, allowing either to be isolated, or the two to be observed (or detected) simultaneously but separately. Also it is clear that, by rotating this prism around the propagation axis, we may reverse the positions of the two components.

It is extremely useful to be able to separate the two orthogonally polarized components in this controllable way.

3.7 Linear Birefringence

The crystal phenomenon we have been considering, whereby two, particular, orthogonal, linear polarization directions (those for which **D** and **E** are parallel) have different velocities, is known as linear birefringence, and those particular polarization states are the linear polarization eigenstates, often known as eigenmodes (this nomenclature is borrowed from the terminology of matrix algebra, for reasons which will become clear in Sections 3.10 and 3.11). These are the only states, in this case, which propagate without change of form (i.e., each of these, as an input state, will emerge at the output also linearly polarized, and in the same polarization direction as the input). Any other input state will change in form: for example, a linear state polarized at some arbitrary angle to these eigenstates directions (the "eigen-axes") will, effectively, be resolved into linear components in the eigenstates' directions, and a phase difference will then be inserted between them, as a result of the velocity difference. The light will, therefore, emerge (in general) elliptically polarized.

We shall now look, briefly, at the other types of birefringence.

3.8 Circular Birefringence

So far we have considered only linear birefringence, where two orthogonal linear polarization eigenstates propagate, each remaining linear, but with different velocities. Some crystals also exhibit circular birefringence. Quartz (again) is one such crystal and its circular birefringence derives from the fact that the crystal structure spirals around the optic axis in a right-handed (dextro-rotatory) or left-handed (laevo-rotatory) sense depending on the crystal specimen: both forms exist in nature.

It is not surprising to find, in view of this knowledge, and our understanding of the easy motions of electrons, that light which is right-hand circularly polarized (clockwise rotation of the tip of the electric vector as viewed by a receiver of the light) will travel faster down the axis of a matching right-hand spiraled crystal structure than left-hand circularly polarized light. We now have circular birefringence: the two circular polarization components propagate without change of form (i.e., they remain circularly polarized) but at different velocities. They are the circular polarization eigenstates for this case.

The term "optical activity" traditionally has been applied to this phenomenon, and it is usually described in terms of the rotation of the polarization direction of a linearly polarized wave as it passes down the optic axis of an "optically active" crystal. This fact is exactly equivalent to the interpretation in terms of circular birefringence, since a linear polarization state can be resolved into two oppositely rotating circular components (Figure 3.13). If these travel at different velocities, a phase difference is inserted between them. As a result of this, when recombined, they again form a resultant which is linearly polarized but rotated with respect to the original direction (Figure 3.13). Hence optical activity is equivalent to circular birefringence.

3.9 Elliptical Birefringence

In general, both linear and circular birefringence might be present simultaneously in a material (such as quartz). In this case the polarization eigenstates which propagate without change of form (and at different velocities) will be elliptical

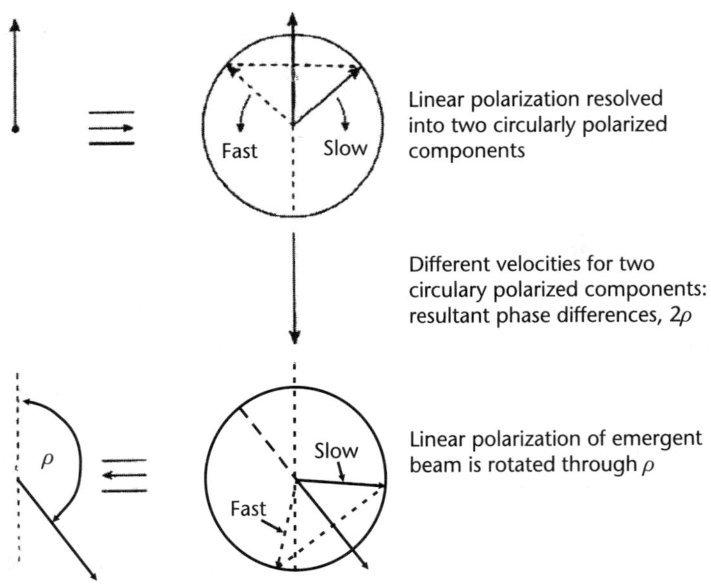

Figure 3.13 Resolution of linear polarization into circularly polarized components in circular birefringence (2ρ).

states, the ellipticity and orientation depending upon the ratio of the magnitudes of the linear and circular birefringences, and on the direction of the linear birefringence eigen-axes within the crystal.

The polarization properties of any lossless, homogeneous, anisotropic element can thus be characterized, at any given optical frequency, by specifying its two normal propagation modes, at that frequency, together with the phase inserted between the modes by the element. These are orthogonal modes, elliptically polarized (in general), which propagate through the element without change of form.

It is often convenient to resolve the polarization behavior of an elliptically birefringent, anisotropic medium into its linear and circular birefringence components, for these can usually be identified with distinct physical mechanisms.

For a medium that exhibits only linear birefringence, the two normal modes are linearly polarized. Thus, we refer, in the specification exercise, to the directions of the fast and slow axes, which correspond to the directions of these two linear modes, and also to the velocity difference that exists between them. The phase difference inserted between the normal modes per unit length (at a given optical wavelength) is designated the linear birefringence, δ. This type of birefringence, as we know, corresponds to a variation with direction of the restrictions imposed on the linear transverse vibrations of the molecular electrons (e.g., unidirectional transverse pressure or transverse electric field).

Correspondingly, for a medium which exhibits only circular birefringence, the two normal modes are circularly polarized, one right-handed the other left-handed. Again, a velocity difference will exist between them and the value of this per unit length is designated 2ρ, being positive if the right-handed rotation is the faster (the factor of 2 is included for convenience because the effect of this birefringence on a polarization ellipse is to rotate it through an angle ρ). This type of birefringence is identifiable physically with restrictions on the rotational motions of the molecular electrons about the longitudinal propagation axis (e.g., spirality in the structure; longitudinal magnetic field).

In general, both effects will be present simultaneously and the normal modes will be elliptically polarized, as indicated earlier. The relationship between the normal mode ellipses and the two birefringence components is readily proved with straightforward ellipse algebra (see Appendix D) and is as follows:

1. The fast and slow linear birefringence axes correspond to the major and minor axes of the two ellipses. The ellipse whose major axis lies in the fast direction will be circumscribed in the same direction as that of the faster of the two circularly polarized components, and vice versa.
2. The ellipticities (i.e., major axis/minor axis) of the two ellipses are given by

$$e = \pm \tan \chi$$

where

$$\tan 2\chi = 2\rho / \delta$$

3. The phase delay, Δ per unit length, inserted by the element between the two ellipses is given by

$$\Delta^2 = (\delta^2 + 4\rho^2)$$

It should, again, be emphasized that only the polarization eigenstates propagate without change of form. All other polarization states will be changed into different polarization states by the action of the polarization element (e.g., a crystal component). These changes of polarization state are very useful in optoelectronics. They allow us to control, analyze, modulate, and demodulate polarization information impressed upon a light beam, and to measure important directional properties relating to the medium through which the light has passed. We must now develop a rigorous formalism to handle these more general polarization processes.

3.10 Polarization Analysis

As has been stated, with both linear and circular birefringence present, the polarization eigenstates (i.e., the states which propagate without change of form) for a given optical element are elliptical states, and the element is said to exhibit elliptical birefringence, since these eigenstates propagate with different velocities.

In general, if we have, as an input to a polarization-optical element, light of one elliptical polarization state, it will be converted, on emergence, into a different elliptical polarization state (the only exceptions being, of course, when the input state is itself an eigenstate). We know that any elliptical polarization state can always be expressed in terms of two orthogonal electric field components defined with respect to chosen axes Ox, Oy; that is,

$$E_x = e_x \cos(\omega t - kz + \delta_x)$$

$$E_y = e_y \cos(\omega t - kz + \delta_y)$$

or, in complex exponential notation:

$$E_x = e_x \exp(i\varphi_x); \qquad \varphi_x = \omega t - kz + \delta_x$$
$$E_y = e_y \exp(i\varphi_y); \qquad \varphi_y = \omega t - kz + \delta_y$$

When this ellipse is converted into another by the action of a lossless polarization element, the new ellipse will be formed from components that are linear combinations of the old, since it results from directional resolutions and rotations of the original fields. Thus these new components can be written:

$$E_x' = m_1 E_x + m_4 E_y$$
$$E_y' = m_3 E_y + m_2 E_y$$

or, in matrix notation:

$$\mathbf{E}' = \mathbf{M} \cdot \mathbf{E}$$

where

$$\mathbf{M} = \begin{pmatrix} m_1 & m_4 \\ m_3 & m_2 \end{pmatrix} \tag{3.2}$$

and the m_n are, in general, complex numbers, to allow for phase changes in the components. **M** is known as a "Jones" matrix after the mathematician who developed an extremely useful *Jones calculus* for manipulations in polarization optics [2]. Now in order to make measurements of the input and output states in practice we need a quick and convenient experimental method. In Section 3.4 there was described a method for doing this which involved the manual rotation of a quarter-wave plate and/or a polarizer, but the method we seek now must lend itself to automatic operation.

A convenient method for this practical determination is again to use the linear polarizer and the quarter-wave plate, but to measure the light intensities for a series of fixed orientations of these elements.

Suppose that $I(\vartheta, \epsilon)$ denotes the intensity of the incident light passed by the linear polarizer set at angle ϑ to Ox, after the Oy component has been retarded by angle ϵ as a result of the insertion of the quarter-wave plate with its axes parallel with Ox, Oy. We measure what are called the four Stokes parameters, as follows:

$$S_0 = I(0°, 0) + I(90°, 0) = e_x^2 + e_y^2$$
$$S_1 = I(0°, 0) - I(90°, 0) = e_x^2 - e_y^2$$
$$S_2 = I(45°, 0) - I(135°, 0) = 2e_x e_y \cos \delta$$
$$S_3 = I\left(45°, \frac{\pi}{2}\right) - I\left(135°, \frac{\pi}{2}\right) = 2e_x e_y \sin \delta$$
$$\delta = \delta_y - \delta_x$$

If the light is 100% polarized, only three of these parameters are independent, since

$$S_0^2 = S_1^2 + S_2^2 + S_3^2$$

S_0 being the total light intensity.

If the light is only partially polarized, the fraction

$$\eta = \frac{S_1^2 + S_2^2 + S_3^2}{S_0^2}$$

defines the degree of polarization. In what follows we shall assume that the light is fully polarized ($\eta = 1$). It is easy to show (see Appendix C) that measurement of the S_n provides the ellipticity, e, and the orientation α of the polarization ellipse according to the relations:

$$e = \tan \chi$$
$$\sin 2\chi = \frac{S_3}{S_0}$$
$$\tan 2\alpha = \frac{S_2}{S_1}$$

Now, the above relations suggest a geometrical construction that provides a powerful and elegant means for description and analysis of polarization-optical phenomena. The Stokes parameters S_1, S_2, S_3 may be regarded as the Cartesian coordinates of a point referred to axes Ox_1, Ox_2, Ox_3. Thus every elliptical polarization state corresponds to a unique point in three-dimensional space. For a constant S_0 (lossless medium) it follows that all such points lie on a sphere of radius S_0—the Poincaré sphere (Figure 3.14). The properties of the sphere are quite well known (see, for example, [3]). We can see that the equator

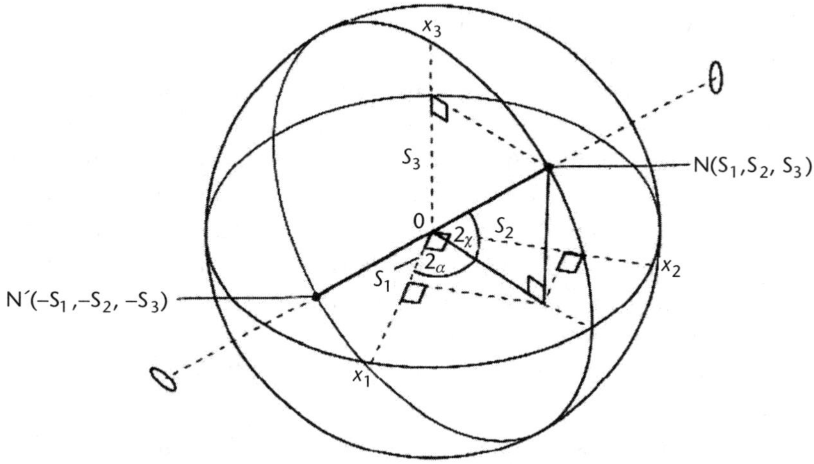

Figure 3.14 The Poincaré sphere: the eigenmode diameter (NN').

will comprise the continuum of linearly polarized states, while the two poles will correspond to the two oppositely handed states of circular polarization.

It is clear that any change, resulting from the passage of light through a lossless element, from one polarization state to another, corresponds to a rotation of the sphere about a diameter. Now any such rotation of the sphere may be expressed as a unitary 2×2 matrix $\mathbf{M}$. Thus, the conversion from one polarization state $\mathbf{E}$ to another $\mathbf{E}'$ may also be expressed in the form:

$$\mathbf{E}' = \mathbf{ME}$$

or

$$\begin{pmatrix} E'_x \\ E'_y \end{pmatrix} = \begin{pmatrix} m_1 & m_4 \\ m_3 & m_2 \end{pmatrix} \begin{pmatrix} E_x \\ E_y \end{pmatrix}$$

that is,

$$E'_x = m_1 E_x + m_4 E_y$$

$$E'_y = m_3 E_y + m_2 E_y$$

where

$$\mathbf{M} = \begin{pmatrix} m_1 & m_4 \\ m_3 & m_2 \end{pmatrix}$$

and $\mathbf{M}$ may be immediately identified with our previous $\mathbf{M}$ in (3.2).

M is a Jones matrix [2] that completely characterizes the polarization action of the element and is also equivalent to a rotation of the Poincaré sphere.

The two eigenvectors of the matrix correspond to the eigenmodes (or eigenstates) of the element (i.e., those polarization states which can propagate through the element without change of form). These two polarization eigenstates lie at opposite ends of a diameter (NN') of the Poincaré sphere and the polarization effect of the element is to rotate the sphere about this diameter (Figure 3.15) through an angle Δ which is equal to the phase which the polarization element inserts between its eigenstates.

The polarization action of the element may thus be regarded as that of resolving the input polarization state into the two eigenstates with appropriate amplitudes, and then inserting a phase difference between them before recombining to obtain the emergent state. Thus, a pure rotator (e.g., optically active crystal) is equivalent to a rotation about the polar axis, with the two oppositely handed circular polarizations as eigenstates. The phase velocity difference between these two eigenstates is a measure of the circular birefringence. Analogously, a pure linear retarder (such as a wave plate) inserts a phase difference between orthogonal linear polarizations that measures the linear birefringence. The linear retarder's eigenstates lie at opposite ends of an equatorial diameter.

It is useful for many purposes to resolve the polarization action of any given element into its linear and circular birefringence components. The Poincaré sphere makes it clear that this may always be done since any rotation of the sphere can always be resolved into two sub-rotations, one about the polar diameter and the other about an equatorial diameter.

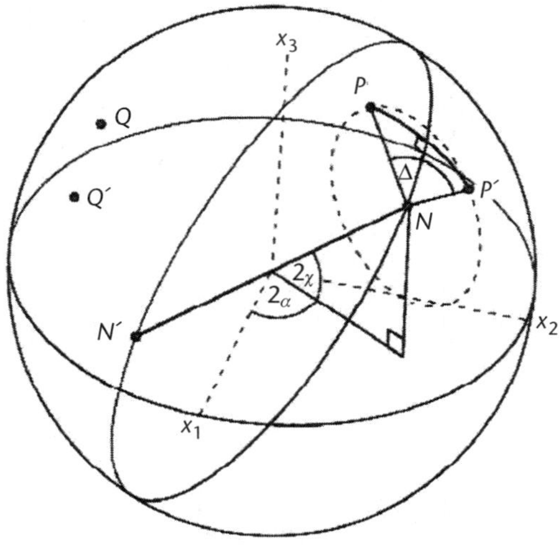

Figure 3.15 Rotation of the Poincaré sphere about the eigenmode diameter NN'.

From this brief discussion we can begin to understand the importance of the Poincaré sphere.

It is a construction that converts all polarization actions into relationships that can be visualized in three-dimensional space.

To illustrate this point graphically let us consider a particular problem. Suppose that we ask what is the smallest number of measurements necessary to define completely the polarization properties of a given lossless polarization element, about which we know nothing in advance. Clearly, we must provide known polarization input states and measure their corresponding output states, but how many input/output pairs are necessary: one, two, more?

The Poincaré sphere answers this question easily. The element in question will possess two polarization eigenmodes and these will be at opposite ends of a diameter. We need to identify this diameter. We know that the action of the element is equivalent to a rotation of the sphere about this diameter, and through an angle equal to the phase difference that the element inserts between its eigenmodes. Hence, if we know one input/output pair of polarization states (NN'), we know that the rotation from the input to the output state must have taken place about a diameter that lies in the plane which perpendicularly bisects the line joining the two states (see Figure 3.15). Two other input/output states (QQ') will similarly define another such plane, and thus the required diameter is clearly seen as the common line of intersection of these planes.

Further, the phase difference Δ inserted between the eigenstates (i.e., the sphere's rotation angle) is easily calculated from either pair of states, once the diameter is known.

Hence the answer is that *two* pairs of input/output states will define completely the polarization properties of the element. Simple geometry has provided the answer. A good general approach is to use the Poincaré sphere to determine (visualize) the nature of the solution to a problem, and then to revert to the Jones matrices to perform the precise calculations. Alternatively, some simple results in spherical trigonometry will usually suffice.

Another important result, more directly relevant to our purposes, also is readily available using the Poincaré sphere. This is the equivalence of any uniform element, possessing distributed, coincident, linear, and circular birefringences, to a series arrangement of two elements, one a retarder the other a rotator. This combination is usually referred to as a retarder/rotator pair (Figure 3.16).

To establish the equivalence, consider again the eigenstates for the uniform birefringent element. Those will be mutually orthogonal ellipses, as we know, and will correspond to points at the opposite ends of a diameter, NN', of the Poincaré sphere. The action of the element on any arbitrary input polarization state, say that at point P (Figure 3.15) will be to rotate the sphere about the diameter NN' through the angle Δ, to yield the final polarization state, P'. Now it is clear from the sphere geometry that the transition from P to P' also

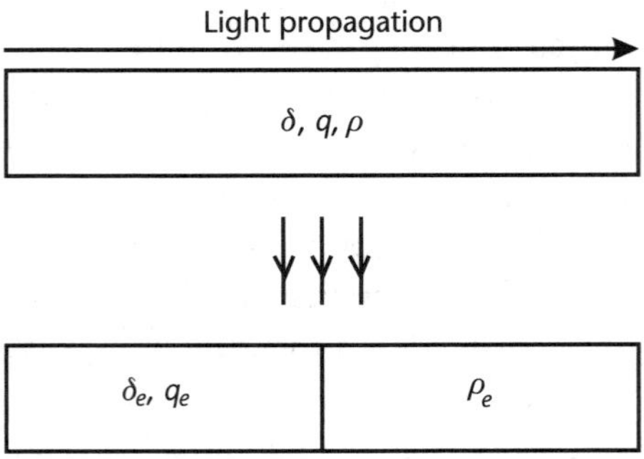

Figure 3.16 Equivalence of the retarder/rotator pair.

can be achieved by a rotation about an equatorial diameter, EE', which lies normal to the great-circle plane containing both OO' and P, to A, followed by one about the polar diameter, OO' (see Figure 3.17) from A to P'. The choice of EE' ensures that the first rotation lies on a great circle, so that a plane normal to OO' and containing P' is always intersected.

Now a rotation about EE' is equivalent to the action of a retarder, because E and E' are linear states, and that about OO' to the action of a circular retarder, or rotator, because O and O' are the two circular states. Hence the required equivalence to a retarder/rotator pair is proved.

We must now study the Jones matrices in more mathematical detail.

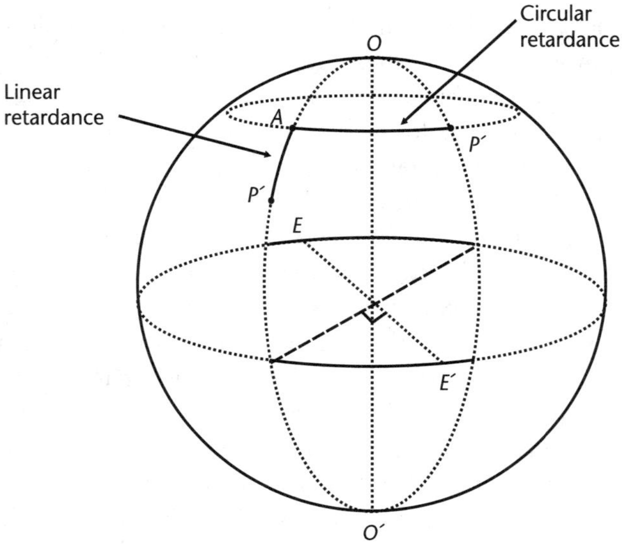

Figure 3.17 Poincaré-sphere representation of the action of a retarder/rotator pair.

3.11 The Form of the Jones Matrices

In order to perform calculations in polarization analysis we need to use the Jones matrix algebra, and we shall now look at this in more detail.

First, we shall assume that the materials in use are all "optical" materials, chosen (amongst other attributes) for their low optical propagation loss. In fact, we shall assume that the loss, for the optical paths under consideration, is negligible. This, certainly, will be the case for monomode optical fibers in current telecommunications usage, which have losses less than 0.2 $dB.km^{-1}$.

Let us first examine the form of the Jones matrices for the two most important types of polarization property: linear and circular birefringence.

3.11.1 Linear Birefringence Matrix

Suppose that we define linearly polarized Cartesian axes *Ox*, *Oy*, *Oz*, and that light propagates in the direction *Oz*. We know that the light can be represented by means of its two (in-phase) electric-field components in the *Ox* and *Oy* directions:

$$E_x = e_x \exp i\omega t$$
$$E_y = e_y \exp i\omega t$$

Suppose now that the x component advances on the y component by a phase angle δ, as a result of passage through a birefringent element with its birefringent axes aligned with *Ox* and *Oy*, *Ox* clearly being (in this case) the fast axis. Then the components after the passage through the element can be written:

$$E'_x = e_x \exp i(\omega t' + \delta/2)$$
$$E'_y = e_y \exp i(\omega t' - \delta/2)$$

where $t' - t = t_0$, say, is the mean time for passage through the element.

It is clear that the input state will be converted into the output state by means of the matrix transformation:

$$\begin{pmatrix} E'_x \\ E'_y \end{pmatrix} = \exp i\omega t_0 \begin{pmatrix} \exp(i\delta/2) & 0 \\ 0 & \exp(-i\delta/2) \end{pmatrix} \begin{pmatrix} E_x \\ E_y \end{pmatrix}$$

Hence the Jones matrix, $\mathbf{M}_l$, for the essential linear birefringence action can be written:

$$\mathbf{M}_1 = \exp i\omega t_0 \begin{pmatrix} \exp(i\delta/2) & 0 \\ 0 & \exp(-i\delta/2) \end{pmatrix} \tag{3.3}$$

and the transformation can be written, compactly:

$$\mathbf{E}' = \mathbf{M}_1\mathbf{E}$$

Note that the elements of the matrix will, in general, be complex, to allow for phase changes during optical passage.

3.11.2 Circular Birefringence Matrix

If the same two components as in Section 3.11.1 are rotated anticlockwise from Ox through an angle ρ by the action of a circularly birefringent element, then the components at the output are given by the standard *rotation of axes* transformation:

$$E_x' = E_x \cos\rho + E_y \sin\rho$$

$$E_y' = -E_x \sin\rho + E_y \cos\rho$$

Hence, in this case,

$$\begin{pmatrix} E_x' \\ E_y' \end{pmatrix} = \exp i\omega t_0 \begin{pmatrix} \cos\rho & \sin\rho \\ -\sin\rho & \cos\rho \end{pmatrix} \begin{pmatrix} E_x \\ E_y \end{pmatrix}$$

or

$$\mathbf{E}' = \mathbf{M}_1\mathbf{E}$$

where

$$\mathbf{M}_1 = \exp i\omega t_0 \begin{pmatrix} \cos\rho & \sin\rho \\ -\sin\rho & \cos\rho \end{pmatrix} \tag{3.4}$$

3.11.3 Elliptical Birefringence Matrix

When both circular and linear birefringences are present together, and uniformly distributed, the calculations are more difficult. Clearly we cannot simply multiply the matrices together because that implies a linear element in series with (e.g., followed by) a circular element, which is not the physical situation. Equally clearly we cannot add the matrices, because then the birefringences would not interact and, physically, it is clear that they do (e.g., when rotation, resulting from circular birefringence, occurs relative to the linear axes, the effect of the linear birefringence is altered).

Jones solves the problem very neatly by using physical insight allied to some well-known results in matrix algebra. We shall not here reproduce all of his detailed calculations but shall merely provide the essence of his argument, to allow the reader to understand the methodology. Of course, the very interested reader can refer to the original papers for a full explanation.

3.11.4 The Essence of the Jones Calculus

For simplicity we shall provide the essence of Jones's solution of the problem for just the two birefringences we have been considering: linear and circular birefringences. From the methodology it will be clear that the argument is extendable to any number of polarization actions (Jones considers eight, including differential linear and circular absorption, for example).

Let us define a set of Cartesian axes, Ox, Oy, Oz, and suppose that we have an element with uniformly-distributed linear and circular birefringences acting simultaneously, and that the light is propagating through it in the Oz direction along a length z_0. Let the linear birefringence be δ per unit length, and the circular birefringence be 2ρ per unit length. To simplify the algebra, we shall firstly assume that the axes Ox, Oy coincide with the fast and slow axes of the linear birefringence.

Let us designate the matrix which we are seeking to evaluate—that is, the Jones polarization matrix of the element—as $\mathbf{M}$. Hence any input polarization vector (E_x, E_y) will be transformed into an output vector (E'_x, E'_y) by the operation

$$\mathbf{E}' = \mathbf{M}\mathbf{E}$$

where the components of E, E' and the elements of M are complex numbers.

The first act of analysis is to define a new matrix, $\mathbf{N}$, given by

$$\mathbf{N} = \frac{d\mathbf{M}}{dz} \cdot \mathbf{M}^{-1} \qquad (3.5)$$

We may integrate this to give

$$\mathbf{M} = \exp(\mathbf{N}z) \qquad (3.6)$$

where the constant of integration has been determined by the the requirement for $\mathbf{M} = \mathbf{I}$ (the identity matrix) at $z = 0$, for no polarization change will occur before the light enters the element. The identity matrix,

$$\mathbf{I} = \begin{pmatrix} 1 & 0 \\ 0 & 1 \end{pmatrix}$$

effects this preservation for any vector on which it acts.

Let us now consider a thin slice of the element, of thickness τ. The polarization matrix for the element can, from (3.6), now be written in the form:

$$\mathbf{M_e} = \exp(\mathbf{N_e}\tau) \tag{3.7}$$

Because $\mathbf{N_e}\tau$ is small, we can expand $\mathbf{M_e}$ as

$$\mathbf{M_e} = \mathbf{1} + \mathbf{N_e}\tau + (\mathbf{N_e}\tau)^2/2! + \ldots \tag{3.8}$$

where the "**1**," in a matrix equation, refers to the identity matrix.

The next step is to subdivide the thin slice into two, of thicknesses τ_1 and τ_2 (i.e., $\tau_1 + \tau_2 = \tau$). If the matrices associated with these new slices are $\mathbf{M_1}$ and $\mathbf{M_2}$, it follows that

$$\mathbf{M_e} = \mathbf{M_2}\mathbf{M_1} \tag{3.9}$$

(Note that $\mathbf{M_1}$ acts first on the vector, and therefore appears after $\mathbf{M_2}$.)

Because $\tau_1, \tau_2 < \tau$, it follows that

$$\begin{aligned} \mathbf{M_1} &= \mathbf{1} + \mathbf{N_1}\tau_1 + \mathbf{O}\left(\tau_1^2\right) \\ \mathbf{M_2} &= \mathbf{1} + \mathbf{N_2}\tau_2 + \mathbf{O}\left(\tau_2^2\right) \end{aligned} \tag{3.10}$$

where $\mathbf{O}(\tau^2)$ represents all subsequent terms of order τ^2 or higher powers; these terms are, of course, negligible compared with $\mathbf{N}\tau$ when τ is very small. So we now have

$$\mathbf{M_e} = \mathbf{M_2}\mathbf{M_1} = \mathbf{1} + \mathbf{N_1}\tau_1 + \mathbf{N_2}\tau_2 + \mathbf{O}(\tau^2) \tag{3.11}$$

Let us now define an average $\mathbf{N}$ matrix by

$$\overline{\mathbf{N}} = (\mathbf{N_1}\tau_1 + \mathbf{N_2}\tau_2)/\tau$$

Then we may write

$$\mathbf{M_e} = \mathbf{1} + \overline{\mathbf{N}}\tau + \mathbf{O}(\tau^2) \tag{3.12}$$

The full element now consists of a series of these thin plates, each with matrix $\mathbf{M_e}$. Hence we may write the matrix of the full element as a product of all of these matrices in the form:

$$\mathbf{M} = \mathbf{M_e}^{z_0/\tau}$$

because there will be z_0/τ identical plates in the element of length z_0.

Hence, from (3.12) we have

$$\mathbf{M} = \mathbf{M_e}^{z_0/\tau} = (\mathbf{1} + \overline{\mathbf{N}}\tau + \mathbf{O}(\tau^2))^{z_0/\tau}$$

For the continuous distribution we now let $\tau \to 0$, and hence we require to evaluate

$$\lim_{\tau \to 0} (1 + \overline{\mathbf{N}}\tau)^{z_0/\tau}$$

Expanding via the binomial theorem and taking to the limit, this can be written:

$$\mathbf{M} = \mathbf{1} + \overline{\mathbf{N}}z_0 + (\overline{\mathbf{N}}z_0)^2/2! + (\overline{\mathbf{N}}z_0)^3/3! + \ldots$$

$$\mathbf{M} = \exp(\overline{\mathbf{N}}z_0) \tag{3.13}$$

So that, from (3.6), $\overline{\mathbf{N}}$ becomes the $\mathbf{N}$ matrix for $\mathbf{M}$.

Suppose, now, that we consider again the two thin subslices of thicknesses τ_1 and τ_2. Let the first of these be a pure retarder, with linear birefringence δ per unit length and axes parallel with *Ox*, *Oy*. From (3.3), if τ is very small, we have

$$\mathbf{M_1} = \begin{pmatrix} 1 + i\delta\tau_1/2 & 0 \\ 0 & 1 - i\delta\tau_1/2 \end{pmatrix}$$

Similarly, let the second slice be a pure rotator with circular birefringence 2ρ per unit length, so that, from (3.4),

$$\mathbf{M_2} = \begin{pmatrix} 1 & \rho\tau_2 \\ -\rho\tau_2 & 1 \end{pmatrix}$$

From (3.10) we see that, in the limit as $\tau \to 0$,

$$\mathbf{N_1}\tau_1 = \mathbf{M_1} - \mathbf{1} = \begin{pmatrix} i\delta\tau_1/2 & 0 \\ 0 & -i\delta\tau_1/2 \end{pmatrix}$$

and

$$\mathbf{N_2}\tau_2 = \mathbf{M_2} - \mathbf{1} = \begin{pmatrix} 0 & \rho\tau_2 \\ -\rho\tau_2 & 0 \end{pmatrix}$$

Hence in this case,

$$\overline{\mathbf{N}} = (\mathbf{N_1}\tau_1 + \mathbf{N_2}\tau_2)/\tau = 1/\tau\begin{pmatrix} i\delta\tau_1/2 & \rho\tau_2 \\ -\rho\tau_2 & -i\delta\tau_1/2 \end{pmatrix}$$

We now effect a final simplification, without loss of generality, by making the thicknesses of the two slices equal; that is,

$$\tau_1 = \tau_2 = \tau/2$$

giving

$$\overline{\mathbf{N}} = 1/2\begin{pmatrix} i\delta/2 & \rho \\ -\rho & -i\delta/2 \end{pmatrix} \tag{3.14}$$

Hence we know the $\overline{\mathbf{N}}$ matrix for the element, so it remains only to calculate the **M** matrix from (3.13). From this point the derivation is a matter of pure matrix algebra.

Jones pusues this derivation by developing the relationships between the eigenvectors and eigenvalues of **M** and $\overline{\mathbf{N}}$, because any matrix can be constructed from the known values of these.

The eigenvectors of a matrix are those vectors which are not changed in direction by the action of the matrix, only in magnitude. And the factors by which the magnitudes are changed are the eigenvalues. In the case of polarization optics, the eigenvector is that polarization state which is unchanged in form (e.g., an ellipse with a given ellipticity and orientation) by the action of the element—the emerging ellipse has the same form as this one which enters.

We first show that the eigenvectors for **M** and $\overline{\mathbf{N}}$ are the same. If the two eigenvectors of **M** are represented by $\mathbf{E_M}$ then, by definition,

$$\mathbf{ME_M} = \lambda_M\mathbf{E_M} \tag{3.15}$$

where the λ_M are the eigenvalues. Differentiating this equation with respect to z:

$$\frac{d\mathbf{M}}{dz} \cdot \mathbf{E_M} = \frac{d\lambda_M}{dz} \cdot \mathbf{E_M} \tag{3.16}$$

(the $\mathbf{E_M}$ are, of course, independent of z).

Now inverting (3.15) we find

$$\mathbf{E_M} = \lambda_M \mathbf{M}^{-1} \mathbf{E_M}$$

where $\mathbf{M}^{-1}$ is the reciprocal of the $\mathbf{M}$ matrix.

Substituting in (3.16) we have

$$\frac{d\mathbf{M}}{dz} \cdot \mathbf{M}^{-1} \lambda_M \mathbf{E_M} = \frac{d\lambda_M}{dz} \cdot \mathbf{E_M}$$

But

$$\frac{d\mathbf{M}}{dz} \cdot \mathbf{M}^{-1} = \overline{\mathbf{N}}$$

from (3.5). Hence,

$$\overline{\mathbf{N}} \mathbf{E_M} = \frac{1}{\lambda_M} \cdot \frac{d\lambda_M}{dz} \cdot \mathbf{E_M} \tag{3.17}$$

Since $\dfrac{1}{\lambda_M} \cdot \dfrac{d\lambda_M}{dz}$ is a scalar, (3.17) is an eigenvector equation for $\overline{\mathbf{N}}$, and thus the $\mathbf{E_M}$ are also the eigenvectors for $\overline{\mathbf{N}}$.

Further, from (3.17), the eigenvalues for $\overline{\mathbf{N}}$ are given by

$$\lambda_N = \frac{1}{\lambda_M} \cdot \frac{d\lambda_M}{dz}$$

or

$$\lambda_M = \exp(\lambda_N z_0) \tag{3.18}$$

where, again, the integration constant has been evaluated by using the fact that $\lambda_M = 1$ when $z = 0$.

So now we evaluate the eigenvectors and eigenvalues of $\overline{\mathbf{N}}$ using (3.14). Then, for **M**, the matrix for the element, the eigenvectors are the same, and **M**'s eigenvalues can be calculated using (3.18). The algebra is straightforward, but tedious, and gives the result:

$$\mathbf{M} = \begin{pmatrix} \alpha + i\beta & -\gamma \\ \gamma & \alpha - i\beta \end{pmatrix} \tag{3.19}$$

where

$$\alpha = \cos \Delta$$

$$\beta = i\delta z_0/2 \cdot \sin \Delta/\Delta$$

$$\gamma = \rho z_0 \cdot \sin \Delta/\Delta$$

with

$$\Delta = z_0(\rho^2 + \delta^2/4)^{1/2}$$

If the linear birefringence fast axis lies at an angle of q (towards Y) with respect to the chosen axes, this matrix generalizes to

$$\mathbf{M} = \begin{pmatrix} \alpha + i\beta \cos 2q & -\gamma + i\beta \sin 2q \\ \gamma + i\beta \sin 2q & \alpha - i\beta \cos 2q \end{pmatrix} \tag{3.20}$$

Hence this is the general matrix for any uniform element possessing only linear and circular birefringence. Its polarization eigenvectors will represent orthogonal ellipses. It should be clear that the method used generalizes to all types of polarization-optical quantities. If there are k of them, then the thin slice is split into k sub-slices, and the averaged **N** value becomes

$$\overline{\mathbf{N}} = \Sigma \mathbf{N_k} \tau_k / \Sigma \tau_k$$

or

$$\overline{\mathbf{N}} = \frac{1}{k} \Sigma \mathbf{N_k}$$

if the k slices all have equal thicknesses.

The rest of the calculation follows the same path as we have taken. Clearly the calculation of the matrix from the resulting eigenvectors and eigenvalues will be more complex. It is for this reason that just two quantities were chosen for easy illustration of the principles of the methodology.

3.11.5 The Retarder/Rotator Pair

In Section 3.10 we noted how the Poincaré sphere could be used to establish the equivalence of a uniform element possessing both linear and circular-birefringence, with a retarding element followed by a rotator—a so-called retarder/rotator pair (Figure 3.16). It is straightforward, from the above analysis, now to establish this equivalence quantitatively by some matrix manipulation.

Suppose that we use the matrix in (3.20) as representing the uniform element, and suppose also that the equivalent retarder (of the pair) has retardance δ_e with orientation q_e, and that the rotation of the equivalent rotator is ρ_e. Then the matrix equation we need to solve is

$$\begin{pmatrix} \cos\rho_e & \sin\rho_e \\ -\sin\rho_e & \cos\rho_e \end{pmatrix}$$
$$\cdot \begin{pmatrix} \cos\delta_e/2 + i\sin\delta_e/2\cdot\cos 2q_e & i\sin\delta_e/2\cdot\sin 2q_e \\ i\sin\delta_e/2\cdot\sin 2q_e & \cos\delta_e/2 - i\sin\delta_e/2\cdot\cos 2q_e \end{pmatrix}$$
$$= \begin{pmatrix} \alpha + i\beta\cos 2q & -\gamma + i\beta\sin 2q \\ \gamma + i\beta\sin 2q & \alpha - i\beta\cos 2q \end{pmatrix}$$

with the values for α, β, γ as in (3.19).

The results of this (cumbersome, but straightforward) calculation give

$$\begin{aligned} \tan\rho_e &= \rho z_0/2\cdot\tan\Delta/\Delta \\ \sin\delta_e/2 &= \delta z_0/2\cdot\sin\Delta/\Delta \\ q_e &= (q - \rho_e/2) \end{aligned} \tag{3.21}$$

Hence, with δ, q, ρ, and z_0 known, δ_e, q_e, and ρ_e can be determined, and the equivalence is quantified.

An extra point of interest, of future relevance (see Chapters 5 and 6), is the situation when polarized light executes a double passage of such a uniform element, first forward and then back. The forward passage through the equivalent linear retarder inserts a retardation and the equivalent rotator then rotates. On backward passage, however, a left-handed rotation (with respect to the propagation direction) becomes a right-handed rotation, and vice versa. Consequently, the rotation is then reversed and cancelled (provided that no magnetic field is acting). The second passage through the retarder doubles its action, because the linear birefringence is independent of propagation direction. The result is that any reciprocal (i.e., no magneto-optic effect) polarization element possessing only linear and circular birefringence will always behave as a pure linear retarder after forward-followed-by-backward passage.

In matrix rotation the result is that, in backward passage, any term varying linearly with ρ is reversed in sign, so that the forward matrix is given by

$$\mathbf{M_F} = \begin{pmatrix} \alpha + i\beta \cos 2q & -\gamma + i\beta \sin 2q \\ \gamma + i\beta \sin 2q & \alpha - i\beta \cos 2q \end{pmatrix}$$

Then the backward matrix becomes

$$\mathbf{M_B} = \begin{pmatrix} \alpha + i\beta \cos 2q & \gamma + i\beta \sin 2q \\ -\gamma + i\beta \sin 2q & \alpha - i\beta \cos 2q \end{pmatrix}$$

that is, the same except for a change of sign for γ because $\gamma = \rho z_0 \sin \Delta/\Delta$. Note that $\Delta = z_0\left(\rho^2 + \frac{\delta^2}{4}\right)^{1/2}$ does not change sign with ρ.

Hence $\mathbf{M_B}$ has the nondiagonal terms of $\mathbf{M_F}$ interchanged. In matrix jargon $\mathbf{M_B}$ is said to be the *transpose* of $\mathbf{M}$ and this is written as

$$\mathbf{M_B} = \overline{\mathbf{M_F}}$$

If we now perform the forward-backward operation on the propagating light, we have

$$\mathbf{E}' = \overline{\mathbf{M_F}} \cdot \mathbf{M_F} \cdot \mathbf{E}$$

Evaluating $\overline{\mathbf{M_F}} \cdot \mathbf{M_F}$, we obtain the matrix of a pure retarder, because the real component of m_3/m_4 vanishes; that is, we find a matrix of the form

$$\begin{pmatrix} a + ib & ic \\ ic & a - ib \end{pmatrix}$$

This manipulation is left as an exercise for the reader.

3.12 Conclusions

In this chapter we have looked closely at the directionality possessed by the optical transverse electric field; that is, we have looked at optical polarization.

We have seen how to describe it, to characterize it, to control it, to analyze it, and how, in some ways, to use it.

We have also looked at the ways in which the transverse electric and magnetic fields interact with directionalities (anisotropies) in material media through which the light propagates. In particular, we first looked at ways in which the interactions allow us to probe the nature and extent of the material directionalities, and thus to understand better the materials themselves.

Second, we looked briefly at the ways in which these material interactions allow us to control light: to modulate it, and perhaps to analyze it.

Lastly, we studied the Jones polarization calculus—a valuable tool which allows calculation of polarization effects, and one of which we shall have much use in later chapters.

We shall see shortly how this knowledge bears directly upon our specific interest in the polarization properties of optical fibers.

References

[1] Nye, J. F., *Physical Properties of Crystals*, Oxford, U.K.: Clarendon Press, 1976, Chapter 2.

[2] Jones, R. C., "A New Calculus for the Treatment of Optical Systems," *J. Opt. Soc. Am.*, Vol. 31 through Vol. 46; 1941 to 1956, pp. 234–241.

[3] Jerrard, H. G., "Transmission of Light Through Optically Active Media," *J. Opt. Soc. Am.*, Vol. 44, No. 8, 1954, pp. 634–664.

Selected Bibliography

Born, M., and E. Wolf, *Principles of Optics*, 5th ed., Oxford, U.K.: Pergamon Press, 1975, Section 1.4.

Collett, E., *Polarized Light: Fundamentals and Applications*, New York: Marcel Dekker, 1993.

Kliger, D. S., J. W. Lewis, and C. E. Randall, *Polarized Light in Optics and Spectroscopy*, Academic Press, 1990.

Shurchiff, W. A., *Polarized Light: Production and Use*, Cambridge, MA: Harvard University Press, 1962 (an excellent introduction).

4

Polarization Effects in Optical Fibers

4.1 Introduction

In this chapter we shall take a look, in general terms, at the physical phenomena which lead to polarization effects in optical fibers. This will give a feel for these effects, which should help when we come to deal more formally with their consequences and applications in the following two chapters.

We have seen that a polarized optical wave is essentially one for which the direction of the transverse electric field is either constant or changing in some ordered, prescribable manner. A medium which effects any change in the polarization state of light which is passing through it, does so by virtue of some directional asymmetry present in the structure of the medium.

We know that when an optical wave passes through a material medium it does so by stimulating the elementary atomic dipoles to radiate. These secondary radiations combine vectorially with the primary wave to give rise to the resultant propagation through the medium, thus defining the latter's (complex) refractive index.

In such circumstances any directionality inherent in the medium itself, resulting either from its crystal structure, from waveguiding geometry, or from externally applied asymmetrical forces, will be impressed also on the propagating wave. Consequently, carefully chosen materials or structures can be used to control polarization state; and the polarization analysis of the resultant wave can be used sensitively to probe material structures.

Clearly, then, polarization effects may arise naturally, or may be induced deliberately. Of those that occur naturally, the most common are the ones that are a consequence of an anisotropic material, an asymmetrical material strain, or asymmetrical waveguide geometries.

If an optical medium is compressed in a particular direction, there results the same kind of directional restriction on the atomic or molecular electrons as in the case of crystals, and hence the optical polarization directions parallel and orthogonal to these imposed forces (for isotropic materials) will encounter different refractive indices. We shall now investigate some of the manifestations of these asymmetries in optical fibers.

4.2 Linear Polarization Effects in Optical Fibers

4.2.1 General Introduction

We begin by examining those polarization effects that are independent of the intensity of the light that is propagating in the fiber. This is, fairly obviously, to distinguish them from those effects that are dependent upon it, a distinction that will become clearer in the next section.

If an optical wave is being guided in a channel, or other type of guide, with a refractive index greater than its surroundings, we have to be aware of the effect of any asymmetry in the geometry of the guide's cross-section. Clearly, if the cross-section is a perfect circle, as in the case of an ideal optical fiber, all linear polarization directions must propagate with the same velocity (provided, of course, that the fiber material itself is homogeneous and isotropic). If, however, the cross-section were elliptical, then it is not difficult to appreciate that a linear polarization direction parallel with the minor axis will propagate at a different velocity from that parallel with the major axis (we should remind ourselves here that, for observable effects with real optical sources, we are referring to the group velocity).

The optical fiber illustrates well these passive polarization effects, since all real fibers possess same directional asymmetry due to one or more of the following: noncircularity of core cross-section; linear strain in the core; twist strain in the core. Bending will introduce linear strain, and twisting will introduce circular strain (Figure 4.1). Linear strain leads to linear birefringence, circular (twist) strain to circular birefringence.

The intrinsic linear birefringence in "standard" telecommunications optical fiber can be quite troublesome for high-performance links since it introduces velocity differences between the two orthogonal linear polarization states, which lead to relative time lags of order 1 to 10 ps.km^{-1}. Clearly, this distorts the modulating signal: a pulse in a digital system, for example, will be broadened, and thus degraded, by this amount. This so-called *polarization mode dispersion* (PMD) can be reduced by spinning the preform from which the fiber is being drawn, while it is being drawn, so as to average out the cross-sectional anisotropies. This *spun preform* technique [1] reduces this form of dispersion to

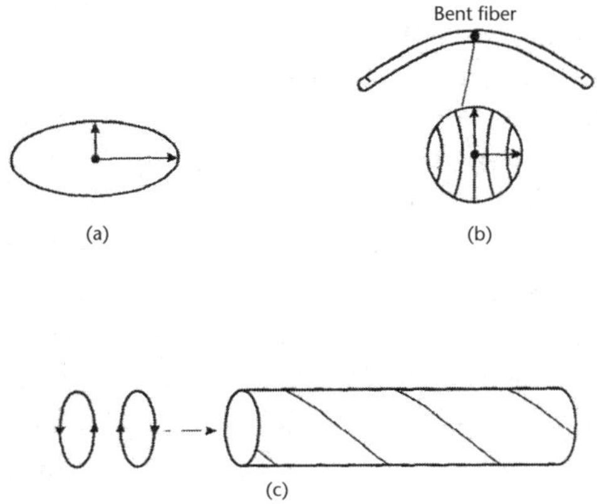

Figure 4.1 Birefringence in optical fibers. Linear birefringenet fibers: (a) geometrical "form"; (b) bending "strain"; and (c) twist-strain circularly birefringent fiber.

~0.01 ps.km^{-1} (i.e., by two orders of magnitude). PMD will be considered in more detail in the next chapter.

It is sometimes valuable to introduce deliberately linear or circular birefringence into a fiber. In order to introduce linear birefringence the fiber core may be made elliptical (with consequences previously discussed) or stress may be introduced by asymmetric doping of the cladding material which surrounds the core (Figure 4.2) [2]. The stress results from asymmetric contraction as the fiber cools from the melt. Circular birefringence may be introduced by twisting and then clamping [Figure 4.1(c)] the fiber. One important application of fiber with a high value of linear birefringence ("hi–bi" fiber) is that linearly polarized light launched into one of the two linear eigenmodes will tend to remain in that state, thus providing a convenient means for conveying linearly polarized light between two points. The reason for this *polarization holding* property is that light, when coupled (i.e., transferred) to the other eigenmode by some random internal or external perturbation, will be coupled to a mode with a different velocity and will not, in general, be in phase with other previous light

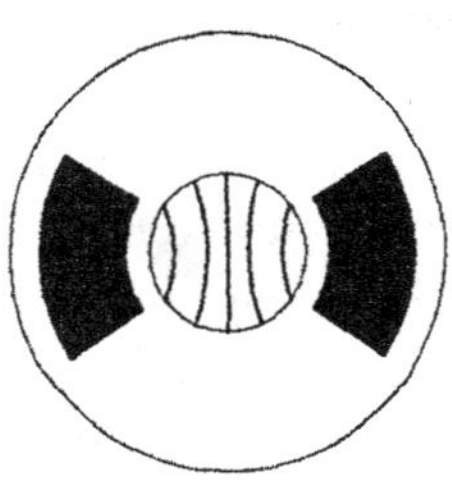

Figure 4.2 Asymmetrically doped linearly birefringent optical fiber ("bow-tie").

couplings into the mode; thus the various couplings will interfere destructively overall and only a small amplitude will result. There is said to be a *phase mismatch*. (This is yet another example of wave interference.) Clearly, however, if a deliberate attempt is made to couple light only at those points where the two modes are in phase, then constructive interference can occur and the coupling will be strong. This is known as *resonant coupling* and has a number of important applications.

The polarization holding also will be resistant to bends and twists in the fiber, provided only that the bend and/or twist-induced birefringence remains small compared with the intrinsically high birefringence of the fiber.

One practical difficulty with hi-bi fiber is that of knowing the orientation of the fast and slow axes when trying, for example, to launch in linearly polarized light, or to join two sections together. A useful development in regard to this was the fabrication of the *D-fiber* [3]. This was an elliptically cored fiber for which the preform has a flat face ground on to one side, corresponding to the plane of the major axis of the core ellipse. This flat face persisted in the drawn fiber, so that the fiber cross-section was roughly that of a D-shape, allowing the axes to be identified readily in practical application.

An extremely convenient way of inducing polarization anisotropies into materials is by subjecting them to (controllable) electric and/or magnetic fields. As we know very well, these fields can exert forces on electrons, so it is not surprising to learn that, via their effects on atomic electrons, the fields can influence the polarization properties of media, just as the chemical-bond restrictions on these electrons in crystals were able to do. The use of electric and magnetic fields thus allows us to build convenient polarization controllers and modulators. Some examples of the effects which can be used will help to establish these ideas.

4.2.2 Polarization-Holding Waveguides

First, consider again the fiber with stress-induced birefringence (two refractive indices). The stress, via the strain-optic effect, has produced a different refractive index for light linearly polarized parallel with the stress direction when compared with that linearly polarized perpendicular to that direction [Figure 4.3(a)]. The difference in refractive indices implies, of course, a difference in velocity between the two polarization states which are, as we know, the two polarization eigenmodes of the guide. This velocity difference means that light of one linear polarization launched into one of the eigenmodes is locked into that state unless the fiber is specially perturbed. The reason for this is that any random couplings from one eigenmode to the other will not, in general, be in phase, owing to the velocity difference between the components. Hence they will tend to interfere destructively. However, coupling can occur if a coupling perturbation—for

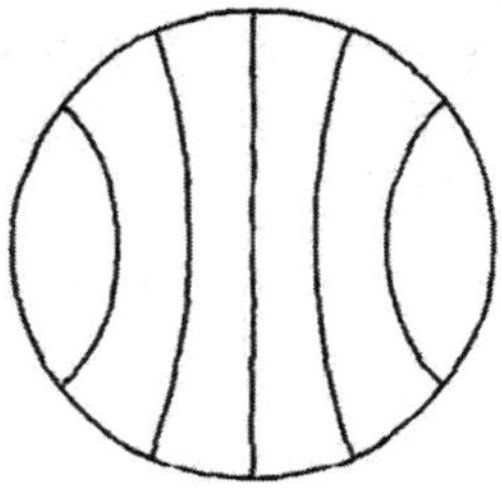

(a)

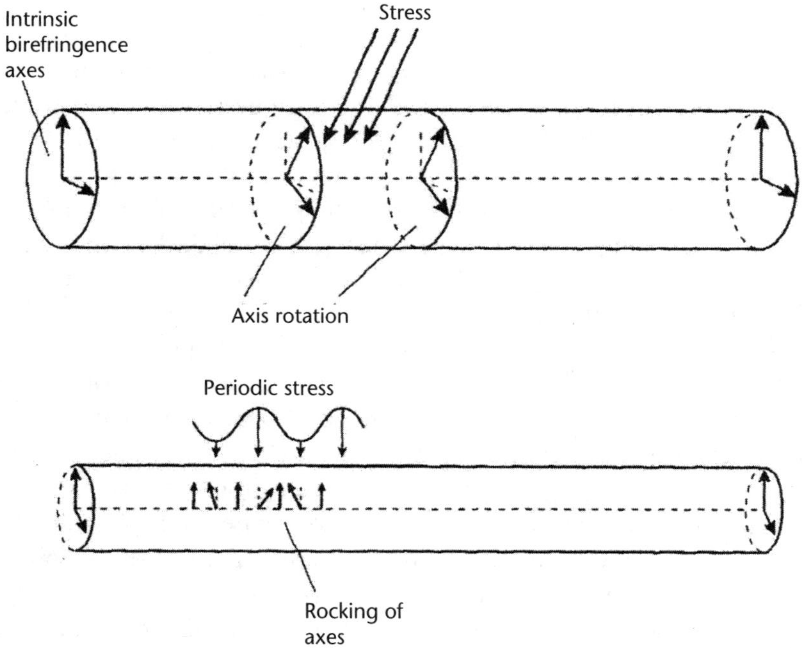

(b)

Figure 4.3 Light coupling in polarization-holding fiber: (a) stress distribution in a polarization-holding fiber; and (b) stress-induced axis rotation.

example, an extra external stress which locally rotates the axes, Figure 4.3(b)—has a spatial period equal to the distance over which the two eigenmodes come into phase. What is this distance? Suppose that the difference in refractive index is Δn; this quantity is also known as the birefringence in this case, and is sometimes given the symbol B. The phase lag due Δn over a distance b is given by

$$\varphi_b = \frac{2\pi}{\lambda} \Delta n \cdot b$$

at wavelength λ. If the two eigenmodes are in phase at a given point in the fiber they will be also in phase after a distance b provided that

$$\varphi_b = \frac{2\pi}{\lambda}\Delta n \cdot b = 2\pi$$

that is,

$$b = \frac{\lambda}{\Delta n} \tag{4.1}$$

The quantity b is known as the *beat length* and is an important characterizing parameter for birefringent fibers. Note that it is proportional to the wavelength of the light. Now, clearly, a perturbation with spatial period equal to b will always couple in-phase components from one eigenmode to the other, so constructive interference will occur, and the coupling is strong. For an arbitrary perturbation function, the coupling will depend upon the amplitude of the Fourier component with the beat length period. The smaller the beat length in a given fiber, the less likely is a given perturbation in practice to contain a large amplitude Fourier component with that small a period. For example, a single weight resting on the fiber would need to apply its stress over a distance of less than one beat length to possess such a component. Consequently, small beat length (i.e., high birefringence) optical fibers will hold a given launched eigenmode very well, and can be used to convey linearly polarized light from one place to another. Such fibers are called *high-birefringence* or, often, more simply *hi-bi* fibers, and are used for polarization control in a variety of applications. They are sometimes also referred to as *polarization-holding fibers* in appropriate applications. A typical hi-bi fiber will have a beat length of ~ 2 mm at a wavelength of 850 nm. From (4.1), this gives

$$\Delta n \sim 4.25 \times 10^{-4}$$

With $n_1 \approx n_2 \sim 1.47$, this is a refractive index difference of only ~ 0.03%, but it is enough to provide strong polarization properties. The basic reason for this is that the short optical wavelengths are operating over much longer optical paths, so that the phase effects quickly accumulate, according to $\varphi = (2\pi/\lambda) \cdot \Delta nl$.

Consider now another type of birefringent fiber, the elliptically cored fiber shown in Figure 4.4. In this case the birefringence is a result of the waveguide geometry, and our recently acquired understanding of waveguiding action allows us some physical insight into the mechanism (the full mathematical description is quite complex and can be studied in [3]).

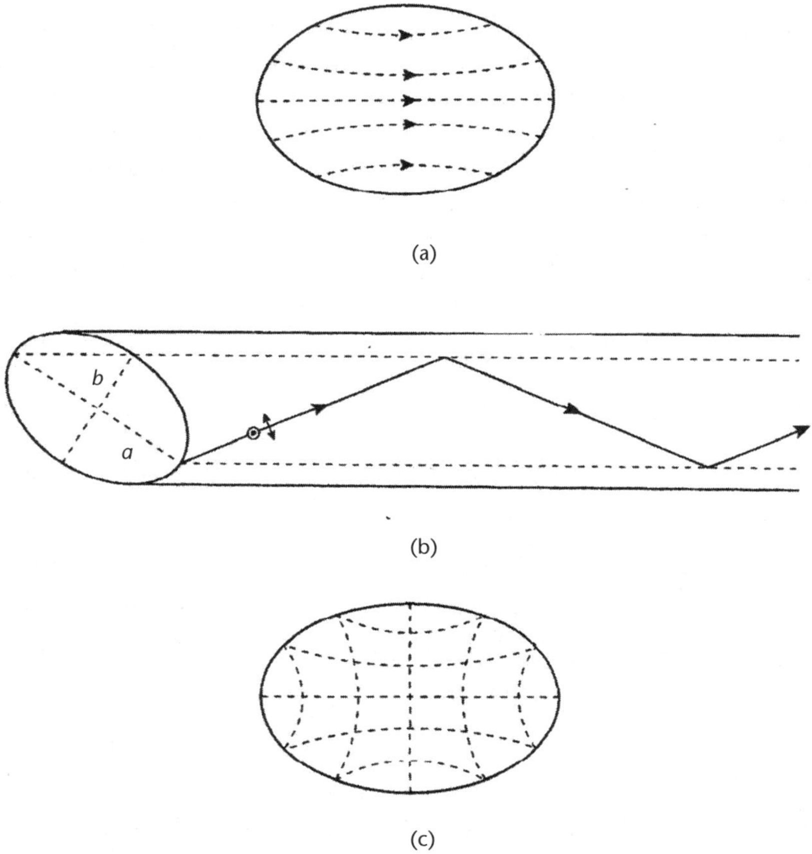

Figure 4.4 The elliptically cored fiber: (a) electrical field distribution for one eigenmode in an elliptically cored fiber; (b) ray diagram for the eigenmode in (a); and (c) E-field distribution for lowest-order orthogonal eigenmodes.

Consider a ray bouncing across the maximum dimension [Figure 4.4(b)] with linear polarization state lying perpendicular to the plane of incidence. Roughly speaking, the condition for constructive interference corresponds to (2.3a) with the major axis equal to a (i.e., $2akn_1 \cos \vartheta + \delta_s = m\pi$).

Hence there is defined a value for ϑ and for δ_s corresponding to that dimension and that polarization state. These parameters, in turn, will define a set of ω/β dispersion curves, like those in Figure 2.3. There will be a different set for the parallel polarization.

For a ray bouncing across the minimum dimension, b, there will be two more such sets, again one for each linear polarization state and, clearly, since $a \neq b$, all four sets will be different. It is now not difficult to appreciate that a combination of any two corresponding perpendicular states leads to a mode which has a different set of ω/β curves from that produced by combining two parallel states [Figure 4.4(c)]. Hence, at a given value of ω (i.e., a given optical frequency), $d\omega/d\beta$ will differ for the two linear polarization states, and we then have differing group velocities, and thus linear birefringence. It is a natural consequence of the symmetry of the ellipse that the directions of the two linear eigenmodes should correspond with those of the axes of the ellipse. It is only when bouncing between major or minor axis extremities that a ray can be confined to one plane (i.e., the plane which contains the guide axis and the ellipse axis). All other rays will criss-cross the ellipse.

It can be shown [3] that, if the ellipticity is not too high and the mode not too far from cut off, the birefringence in this case is given by

$$B \approx 0.28\left(\frac{a}{b} - 1\right)(n_1 - n_2)^2 \tag{4.2}$$

where n_1 is again the core refractive index and n_2 that of the cladding.

Values of a/b usually are in the region of 1.2 for these fibers. The value of $n_1 - n_2$ is limited by the necessity to maintain the single-mode condition:

$$V = \frac{2\pi a,\ b}{\lambda}\left(n_1^2 - n_2^2\right)^{1/2} < 2.405$$

In practice, $n_1^2 - n_2^2$ is made as high as possible, while maintaining the single-mode condition by reducing the dimensions, a and b, of the ellipse. However, this has the effect, of course, of reducing the core area and hence making alignment with sources, and other sections of fiber, rather difficult for these fibers. Another problem is that if n_1 is raised, by too much core doping, the fiber attenuation will rise.

With differences in refractive index between core and cladding (usually by doping the core with germanium) of 4%, (4.1) shows that beat lengths of the order of 1 mm, at ~ 850 nm, can be obtained with such fiber.

One important advantage possessed by elliptically cored hi-bi fiber over the stress-induced variety is that the former has a lower temperature coefficient of birefringence (by a factor ~ 5), owing to the fact that it is geometrically induced, and does not suffer from the temperature dependence of the strain-optic coefficient, as does the stress-induced birefringence.

As was noted in Section 4.2.1, birefringence in optical fibers has disadvantages in optical telecommunications in that it introduces yet another form of dispersion, and hence bandwidth limitation, unless steps are taken to reduce it to acceptable levels, such as spinning the preform from which the fiber is drawn, so as to average out any cross-sectional anisotropies in the waveguide.

Furthermore, linear birefringence is not the only birefringent form that can be present (again as was noted in Section 4.2.1). Any waveguide might possess polarization eigenmodes of any elliptical form, including the linear and circular ones as special cases. These eigenmodes (by definition) will possess different velocities, thus providing, in general, elliptical birefringence in the guide. The effects of such a waveguide on any, arbitrary, input polarization state can be determined analytically by resolving the input state into the two eigenmode components (always possible) and then recombining them at the guide's output with their relative phase equal to that inserted by the guide, as a consequence of the velocity difference of the eigenmodes. In practice, linear and circular birefringences are the types most commonly encountered, and optical fibers of both these varieties are available commercially.

The control that the waveguide allows over the direction and localization of the propagating light is further enhanced by this control over its polarization state, and there are many practical examples in optoelectronics where this control is used to considerable advantage. Some of these applications are described in Chapter 6.

4.2.3 Bend-Induced Linear Birefringence

When an optical fiber is bent, the fused quartz material is strained. The strain, clearly, alters the atomic structure to some small extent and, as a result, the refractive index of the material is changed.

Bends are, by their very nature, anisotropic in their effects: a bend occurs in a particular plane, implying directionality. This directionality itself implies anisotropy in the strain-induced refractive index, and thus birefringence.

For a fiber bent uniformly with bend radius R, the stress/strain effects will differ for components in the plane of the bend compared with those perpendicular to it. The refractive index for light linearly polarized in these two planes also will differ, leading to linear birefringence with axes orientated in these directions.

An analysis of the strain effect for a quartz fiber bent in this way shows that the induced linear birefringence (difference in refractive index for light polarized parallel to and normal to, respectively, the plane of bending) is given by [4]

$$\Delta n = -0.135 r^2 / R^2$$

where r is the outer radius of the fiber and R is the bend radius. To give a feel for magnitudes; for a fiber of radius 50 μm and a bend radius of 50 mm, this expression gives $\Delta n = 1.35 \times 10^{-7}$; at a wavelength of 1,550 nm this implies a beat-length of ~ 10m.

4.2.4 Twist-Induced Circular Birefringence

When a fiber is twisted it is shear-strained. The twist is either right- or left-handed, implying intrinsic rotatory asymmetry and hence circular birefringence. A calculation similar to that of the preceding section shows that this circular birefringence will rotate a linearly polarized wave in the same direction as the twist by an amount [5]:

$$\rho \simeq 0.08\tau$$

where τ is the twist angle. We should also remember that the corresponding circular birefringence is 2ρ.

4.2.5 Twisted Linearly Birefringent Fiber

When a linearly birefringent fiber is twisted the situation is more complicated than in the previous two sections because the linear birefringence axes are also rotating. This, in itself, gives rise to a component of circular birefringence, in addition to that produced by the twist-induced strain.

When the linear birefringence, δ, is much greater than the twist, τ, the resulting effect (to a first order approximation) is simply to rotate the polarization state with the twist, because the twist is incapable of coupling any significant amount of light from one birefringence axis to the other in the face of the polarization-holding property of the linear birefringence. Hence the medium becomes a weak polarization rotator.

When the reverse is true (i.e., $\delta \ll \tau$), the strong twist will give rise to a large strain-induced circular birefringence which will swamp the linear birefringence, essentially converting it, again, into a polarization-rotating medium, this time a strong one.

When $\delta \sim \tau$, a full analysis required, using the Jones calculus. A full analysis requires detailed matrix algebra and is beyond the spirit of this book (which concentrates on physical mechanism) but we can give a flavor of the calculation in order to illustrate the methodology.

Suppose that we have two identical, equal lengths of fiber, each with linear birefringence δ, orientation of birefringence axes $q/2$ (with respect to an arbitrary reference direction) and circular birefringence 2ρ.

From (3.20) we know that each length of fiber can be represented, in regard to its polarization properties, as the matrix:

$$\mathbf{M} = \begin{pmatrix} \alpha + i\beta \cos q & -\gamma + i\beta \sin q \\ \gamma + i\beta \sin q & \alpha - i\beta \cos q \end{pmatrix}$$

with

$$\alpha = \cos \Delta$$
$$\beta = \delta/2 \cdot \sin \Delta/\Delta$$
$$\gamma = \rho \cdot \sin \Delta/\Delta$$
$$\Delta = (\rho^2 + \delta^2/4)^{1/2}$$

Suppose, now, that we rotate the second length through a small angle $\Delta q/2$. Then the polarization effect of the two lengths in series will be given by the product of the two matrices:

$$\mathbf{M}' = \begin{pmatrix} \alpha + i\beta \cos(q + \Delta q) & -\gamma + i\beta \sin(q + \Delta q) \\ \gamma + i\beta \sin(q + \Delta q) & \alpha - i\beta \cos(q + \Delta q) \end{pmatrix} \cdot \begin{pmatrix} \alpha + i\beta \cos q & -\gamma + i\beta \sin q \\ \gamma + i\beta \sin q & \alpha - i\beta \cos q \end{pmatrix}$$

Let us just calculate element m_4 in $\mathbf{M}'$:

$$(\alpha + i\beta \cos(q + \Delta q))(-\gamma + i\beta \sin q) + (-\gamma + i\beta \sin(q + \Delta q))(\alpha - i\beta \cos q)$$

The real terms in this expression represent the circular birefringence elements. Pulling out just these real terms we have

$$-2\alpha\gamma + \beta^2 \sin \Delta q$$

The first term in this expression can be written:

$$-2\alpha\gamma = -2\rho \cdot \cos \Delta \cdot \sin \Delta/\Delta = -2\rho \cdot \sin 2\Delta/2\Delta$$

This represents just a doubling of the effect of the circular birefringence, 2ρ, of each length, as we would expect. The second term, however, $\beta^2 \sin \Delta q$, represents a circular birefringence component that has been added by the twist, Δq.

Clearly, in a real fiber, the rotation would also cause a twist strain, so that we should also have to add a circular birefringence component proportional to this. Of course, for a proper analysis we would need to adopt a continuous approach to the problem rather than the discrete approach used illustratively above. The full approach evidently requires good familiarity with the manipulations of the Jones calculus.

4.2.6 The Electro-Optic Effect

When an electric field is applied to an optical medium, the electrons suffer restricted motion in the direction of the field, when compared with that orthogonal to it. Thus the material becomes linearly birefringent in response to the field. This is known as the electro-optic effect.

Consider the arrangement of Figure 4.5. Here we have incident light which is linearly polarized at 45° to an electric field and the field acts on a medium transversely to the propagation direction of the light. The field-induced linear birefringence will cause a phase displacement between components of the incident light which lie, respectively, parallel and orthogonal to the field; hence the light will emerge elliptically polarized.

A (perfect) polarizer placed with its acceptance direction parallel with the input polarization direction will of course, pass all the light in the absence of a field. When the field is applied, the fraction of light power which is passed will depend upon the form of the ellipse, which in turn depends upon the phase delay introduced by the field. Consequently, the field can be used to modulate the intensity of the light, and the electro-optic effect is, indeed, very useful for the modulation of light.

The phase delay introduced may be proportional either to the field (Pockels effect) or to the square of the field (Kerr effect). All materials manifest a transverse Kerr effect. Only crystalline materials can manifest any kind of Pockels effect, or longitudinal (E field parallel with propagation direction) Kerr effect. This is normally expressed in terms of the associated refractive index change:

$$\Delta n = \mathrm{P} \cdot \mathrm{E} \text{ (Pockels)}$$

$$\Delta n = \mathrm{K} \cdot \mathrm{E}^2 \text{ (Kerr)}$$

where P and K are the Pockels and Kerr coefficients, respectively.

The reason for the crystalline effects is physically quite clear. If a material is to respond linearly to an electric field, the effect of the field must change

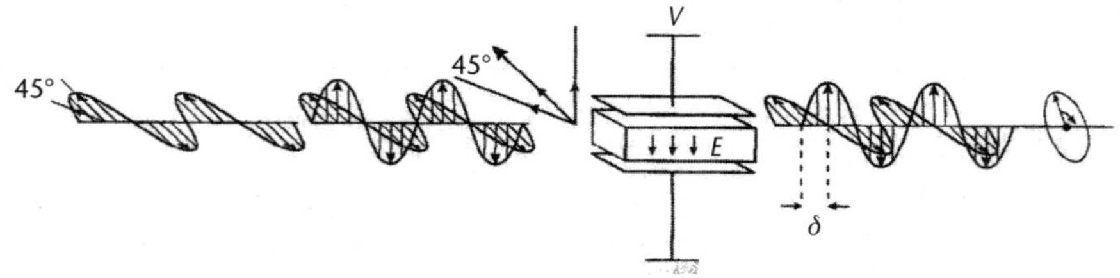

Figure 4.5 The electro-optic effect. Linear polarization becomes elliptical by passing through an electro-optic medium with applied field E.

sign when the field changes sign. This means that the medium must be able to distinguish (for example) between "up" (positive field) and "down" (negative field). But it can only do this if it possesses some kind of directionality in itself, otherwise all field directions must be equivalent in their physical effects. Hence, in order to make the necessary distinction between up and down, the material must possess an intrinsic asymmetry, and hence must be crystalline (however, not all crystals are asymmetric, so the Pockels effect is not present in all). By a similar argument a longitudinal E field can only produce a directional effect orthogonally to itself (i.e., in the direction of the optical electric field) if the medium is anisotropic (i.e., crystalline), for otherwise, all transverse directions will be equivalent. In addition to the modulation of light (phase or intensity/power) it is clear that the electro-optic effect could be used to measure an electric field and/or the voltage which gives rise to it. Several modulators and sensors are based on this idea.

4.2.7 The Magneto-Optic Effect

If a magnetic field is applied to a medium in a direction parallel with the direction in which light is passing through the medium, the result is a rotation of the polarization direction of whatever is the light's polarization state: in general, the polarization ellipse is rotated. The phenomenon, known as the Faraday (after its discoverer, in 1845) magneto-optic effect, normally is used with a linearly polarized input, so that there is a straightforward rotation of a single polarization direction (Figure 4.6). The magnitude of the rotation due to a field, H, over a path length, L, is given by

$$\rho = V \int_0^L H \cdot dl$$

where V is a constant known as the Verdet constant: V is a constant for any given material, but is wavelength and temperature dependent. Clearly, if H is constant over the optical path, we have

$$\rho = VHL$$

From the discussion in Section 3.7, we see that this is a magnetic-field-induced circular birefringence.

The physical reason for the effect is easy to understand in qualitative terms. When a magnetic field is applied to a medium, the atomic electrons find it easier to rotate in one direction around the field than in the other: the Lorentz force acts on a moving charge in a magnetic field, and this will act radially on the electron as it circles the field. The force will be outward for one direction

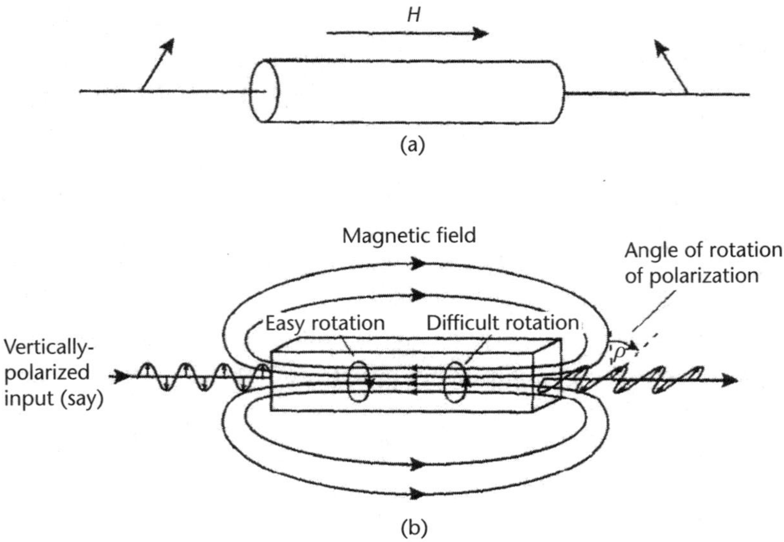

Figure 4.6 The Faraday magneto-optic effect.

of rotation and inward for the other. The consequent electron displacement will lead to two different radii of rotation and thus two different rotational frequencies; and electric permittivities. Hence the field will result in two different refractive indices, and thus to circular birefringence. Light which is circularly polarized in the "easy" (say, clockwise) direction will travel faster than that polarized in the "hard" direction (anticlockwise), leading to the observed effect [Figure 4.6(b)]. Another important aspect of the Faraday magneto-optic effect is that it is *nonreciprocal*. This means that linearly polarized light (for example) is always rotated in the same absolute direction in space, independently of the direction of propagation of the light [Figure 4.7(a)]. For an optically active crystal this is not the case: if the polarization direction is rotated from right to left (say) on forward passage (as viewed by a fixed observer) it will be rotated from left to right on backward passage (as viewed by the same observer), so that back-reflection of light through an optically active crystal will result in light with zero final rotation, the two rotations having cancelled out [Figure 4.7(b)]. This is because the rotation is a result of a longitudinal spirality in the crystal structure. Hence, a rotation following the handedness of the spiral in the forward direction will be opposed by the spiral in the backward direction [Figure 4.7(c)].

For the Faraday magneto-optic case, however, the rotation always takes place in the same direction with respect to the magnetic field (not the propagation direction) since it is this which determines easy and hard directions. Hence, an observer always looking in the direction of light propagation will see different

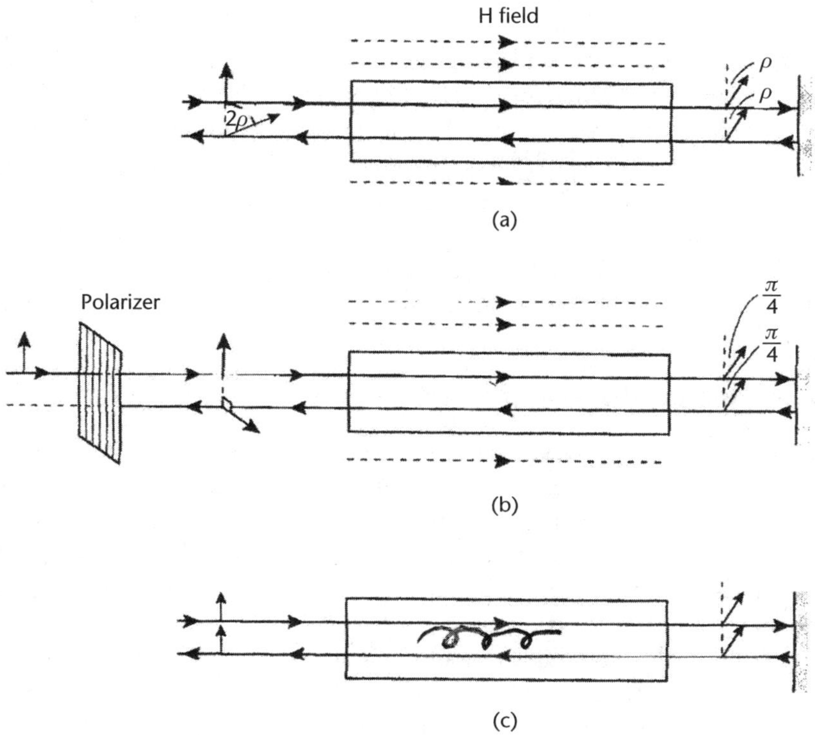

Figure 4.7 Reciprocal and nonreciprocal polarization rotation: (a) nonreciprocal rotation (Faraday effect)—rotation in same direction in relation to the magnetic field; (b) optical isolator action—total rotation of $\pi/2$ for polarization blocking; and (c) reciprocal rotation (optical activity)—rotation in same direction in relation to propagation direction.

directions of rotation since he/she is, in one case, looking along the field and, in the other, against it. It is a nonreciprocal effect. The Faraday effect has a number of practical applications. It can be used to modulate light, although it is less convenient for this than the electro-optic effect, owing to the greater difficulty of producing and manipulating large, and rapidly varying (for high modulation bandwidth) magnetic fields when compared with electric fields (large solenoids have large inductance!).

It can very effectively be used in optical isolators, however. In these devices light from a source passes through a linear polarizer and then through a magneto-optic element that rotates the polarization direction through 45°. Any light which is back-reflected by the ensuing optical system suffers a further 45° rotation during the backward passage, and in the same rotational direction, thus arriving back at the polarizer rotated through 90°; it is thus blocked by the polarizer [Figure 4.5(b)]. Hence the source is isolated from back-reflections by the magneto-optic element/polarizer combination that is thus known as a Faraday magneto-optic isolator. This is very valuable for use with devices whose stability is sensitive to back-reflection, such as lasers and optical amplifiers, and it effectively protects them from feedback effects. The Faraday magneto-optic effect also can be used to measure magnetic fields, and the electric currents that give rise to them.

There are other magneto-optic effects (e.g., Kerr, Cotton-Mouton, Voigt) but the Faraday effect is by far the most important for photonics applications.

4.2.8 Polarization-Dependent Loss/Gain

We know that an optical wave propagates in an optical medium via its interaction with the elementary dipole oscillators comprised by the molecules, and that this interaction is of the nature of the driving-force/natural-resonant-oscillator variety. This approach elucidates the physics of the complex refractive index, its real and imaginary parts representing the effects on phase velocity and attenuation, respectively, as explained in Section 1.7. The ideas can readily be generalized to include amplification of the wave (in addition to its attenuation), if the elementary oscillators are allowed to be in an excited state as a result of being previously stimulated ("pumped") by an external agency.

It is easy, then, to appreciate that, for molecules with any asymmetry, there is likely to be a variation in the absorption or amplification of the wave with polarization state, and thus a dependence of the amplitude of the wave on polarization state, in addition to that on the velocity. This will lead to a polarization-dependent loss or gain. There will be other differential amplitude effects owing to waveguide asymmetries: for example, the evanescent fields for differing polarizations will have differing diameters and will, perhaps, encounter a larger number of absorbing molecules.

Polarization-dependent loss/gain (PDL/G) is characterized by a parameter which gives a measure (usually in decibels) of the ratio of the maximum to minimum signal power for all possible polarization states.

4.3 Nonlinear Polarization Effects in Optical Fibers

4.3.1 General Introduction

In all of the various discussions concerning the propagation of light in material media so far, we have been dealing with linear processes. By this we mean that a light beam of a certain optical frequency that enters a given medium will leave the medium with the same frequency, although the amplitude and phase of the wave will, in general, be altered.

The fundamental physical reason for this linearity lies in the way in which the wave propagates through a material medium. The effect of the electric field of the optical wave on the medium is to set the electrons of the atoms (of which the medium is composed) into forced oscillation; these oscillating electrons then radiate secondary wavelets (since all accelerating electrons radiate) and the secondary wavelets combined with each other and with the original (primary) wave, to form a resultant wave. Now the important point here is that all the forced electrons oscillate at the same frequency (but differing phase, in general) as the primary, driving wave, and thus we have the sum of waves all of the same frequency, but with different amplitudes and phases.

If two such sinusoids are added together:

$$A_T = a_1 \sin(\omega t + \varphi_1) + a_2 \sin(\omega t + \varphi_2)$$

and we have, from simple trigonometry:

$$A_T = a_T \sin(\omega t + \varphi_T)$$

where

$$a_T^2 = a_1^2 + a_2^2 + 2a_1 a_2 \cos(\varphi_1 - \varphi_2)$$

and

$$\tan \varphi_T = \frac{a_1 \sin \varphi_1 + a_2 \sin \varphi_2}{a_1 \cos \varphi_1 + a_2 \cos \varphi_2}$$

In other words, the resultant is a sinusoid of the same frequency but of different amplitude and phase. It follows, then, that no matter how many more such waves are added, the resultant will always be a wave of the same frequency; that is,

$$A_T = \sum_{n=0}^{N} a_n \sin(\omega t + \varphi_n) = \alpha \sin(\omega t + \beta)$$

where α and β are expressible in terms of the a_n and φ_n.

It follows, furthermore, that if there are two primary input waves, each will have the effect described above independently of the other, for each of the driving forces will act independently and the two will add to produce a vector resultant. We call this the *principle of superposition* for linear systems since the resultant effect of the two (or more) actions is just the sum of the effects of each on acting on its own. This has to be the case while the displacements of the electrons from their equilibrium positions in the atoms vary linearly with the force of the optical electric fields. Thus, if we pass into a medium, along the same path, two light waves, of angular frequencies ω_1 and ω_2, emerging from the medium will be two light waves (and only two) with those same frequencies, but with different amplitudes and phases from the input waves.

Suppose now, however, that the displacement of the electrons is *not* linear with the driving force. Suppose, for example, that the displacement is so large that the electron is coming close to the point of breaking free from the atom altogether. We are now in a nonlinear regime. Strange things happen here. For example, a given optical frequency input into the medium may give rise to waves of several different frequencies at the output. Two frequencies ω_1 and ω_2 passing in may lead to, among others, sum and difference frequencies $\omega_1 \pm \omega_2$ coming out.

The fundamental reason for this is that the driving sinusoid has caused the atomic electrons to oscillate nonsinusoidally (Figure 4.8). Our knowledge of Fourier analysis tells us that any periodic nonsinusoidal function contains, in addition to the fundamental component, components at harmonic frequencies (i.e., integral multiples of the fundamental frequency).

This is a fascinating regime. All kinds of interesting new optical phenomena occur here. As might be expected, some are desirable; some are not. Some are valuable in new applications, some just comprise sources of noise. But in order to use them to advantage, and to minimize their effects when they are a nuisance, we must, of course, understand them better. This we shall now try to do.

4.3.2 The Formalism of Nonlinear Optics

Normally we assume a linear relationship between the separation of the positive and negative electric charges, specified quantitatively by the electric polarization

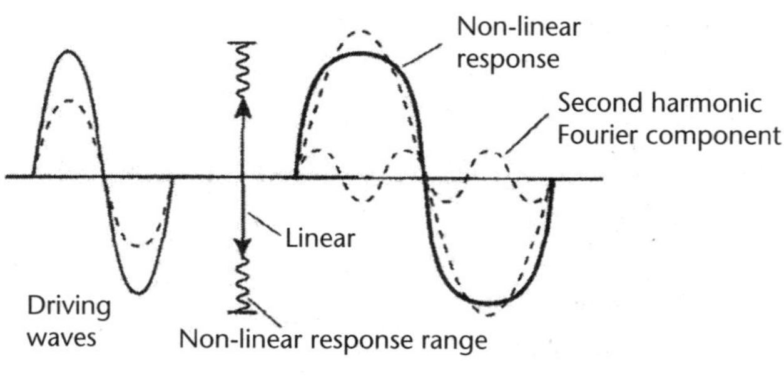

Figure 4.8 Nonlinear response to a sinusoidal drive.

(P), of a medium and the electric field (E) of an optical wave propagating in it, by taking

$$\chi = \frac{P}{E}$$

(for convenience the constant ϵ_0 has been absorbed into χ; this only implies a change of units) where χ is the volume susceptibility of the medium, and is assumed constant. The underlying assumption for this is that the separation of atomic positive and negative charges is proportional to the imposed field, leading to a dipole moment per unit volume (P), which is proportional to the field.

However, it is clear that the linearity of this relationship cannot persist for ever-increasing strengths of field. Any resonant physical system must eventually be torn apart by a sufficiently strong perturbing force and, well before such a catastrophe occurs, we expect the separation of oscillating components to vary nonlinearly with the force. In the case of an atomic system under the influence of the electric field of an optical wave, we can allow for this nonlinear behavior by writing the electric polarization of the medium in the more general form:

$$P(E) = \chi_1 E + \chi_2 E^2 + \chi_3 E^3 + \ldots + \chi_j E^j + \ldots \tag{4.3}$$

The value of χ_j (often written $\chi^{(j)}$) decreases rapidly with increasing j for most materials. Also the importance of the jth term, compared with the first, varies as $(\chi_j/\chi_1)E^{(j-1)}$, and so depends strongly on E. In practice, only the first three terms are of any great importance, and then only for laser-like intensities, with their large electric fields. It is not until one is dealing with power densities of ~ 10^9 Wm^{-2}, and fields ~ 10^6 Vm^{-1}, that $\chi_2 E^2$ becomes comparable with $\chi_1 E$.

Let us now consider the refractive index, n, of the medium. From elementary electromagnetics we know that

$$\epsilon = 1 + \chi,\ n^2 = \epsilon$$

where ϵ is the dielectric constant.

Hence,

$$n = (1 + \chi)^{1/2} = \left(1 + \frac{P}{E}\right)^{1/2}$$

that is,

$$n = \left(1 + \chi_1 + \chi_2 E + \ldots \chi_j E^{j-1} + \ldots\right)^{1/2} \qquad (4.4)$$

Hence we note that the refractive index has become dependent on E. The optical wave, in this nonlinear regime, is altering its own propagation conditions as it travels. This is a central feature of nonlinear optics.

4.3.3 Nonlinear Effects in Optical Fibers

Let us begin by summarizing the conditions that give rise to optical nonlinearity.

In a semiclassical description of light propagation in dielectric media, the optical electric field drives the atomic/molecular oscillators of which the material is composed, and these oscillators become secondary radiators of the field; the primary and secondary fields then combine vectorially to form the resultant wave. The phase of this wave (being different from that of its primary) determines a velocity of light different from that of free space, and its amplitude determines a scattering/absorption coefficient for the material.

Nonlinear behavior occurs when the secondary oscillators are driven beyond the linear response; as a result, the oscillations become nonsinusoidal. Fourier theory dictates that, under these conditions, frequencies other than that of the primary wave will be generated (Figure 4.8).

The fields necessary to do this depend upon the structure of the material, since this is what dictates the allowable range of sinusoidal oscillation at given frequencies. Clearly, it is easier to generate large amplitudes of oscillation when the optical frequencies are close to natural resonances, and one expects (and obtains) enhanced nonlinearity there. The electric field required to produce nonlinearity in material therefore varies widely, from $\sim 10^6\ \mathrm{V \cdot m^{-1}}$ up to $\sim 10^{11}\ \mathrm{V \cdot m^{-1}}$, the latter being comparable with the atomic electric field. Even the lower of these figures, however, corresponds to an optical intensity of $\sim 10^9\ \mathrm{W \cdot m^{-2}}$, which is only achievable practically with laser sources. It is for this reason that the study of nonlinear optics only really began with the invention of the laser, in 1960.

The magnitude of any given nonlinear effect will depend upon the optical intensity, the optical path over which the intensity can be maintained, and the size of the coefficient which characterizes the effect.

In bulk media the magnitude of any nonlinearity is limited by diffraction effects. For a beam of power P watts and wavelength λ focused to a spot of radius r, the intensity, $P/\pi r^2$, can be maintained (to within a factor of ~ 2) over a distance $\sim r^2/\lambda$ (Rayleigh distance), beyond which diffraction will rapidly reduce it. Hence the product of intensity and distance is $\sim P/\pi\lambda$, independent of r, and of propagation length [Figure 4.9(a)].

However, in an optical fiber the waveguiding properties, in a small diameter core, serve to maintain a high intensity over lengths of up to several kilometers

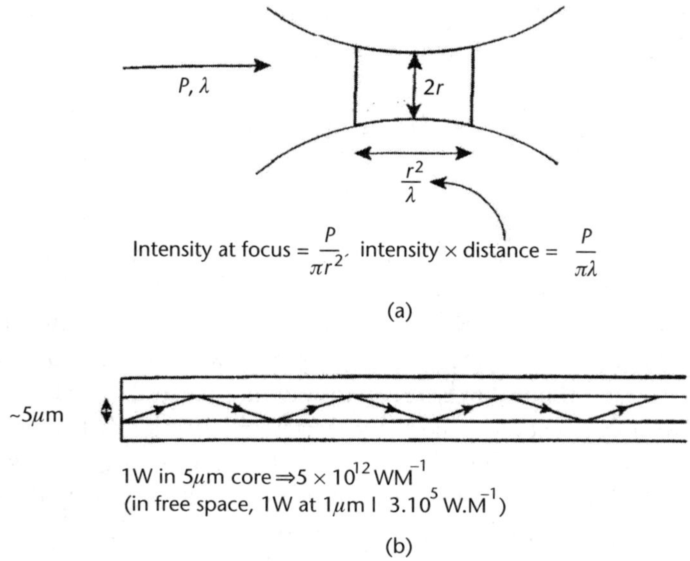

Figure 4.9 The intensity-distance product for nonlinearity: (a) the Rayleigh distance for free-space focusing; and (b) nonlinear facility in optical fibers.

[Figure 4.9(b)]. This simple fact allows magnitudes of nonlinearities, in fibers, which are many orders greater than in bulk materials. Further, for maximum overall effect, the various components' effects per elemental propagation distance must add coherently over the total path. This implies a requirement for phase coherence throughout the path, which, in turn, implies a single propagation mode: monomode rather than multimode fibers must, in general, be used.

4.3.4 Second Harmonic Generation and Phase Matching

Probably the most straightforward consequence of nonlinear optical behavior in a medium is that of the generation of the second harmonic of a fundamental optical frequency. This will be treated in some detail because it illustrates well the important phenomenon of phase matching in polarization-coupling processes: these are especially important for optical fibers.

To appreciate this phenomenon mathematically, let us assume that the electric polarization of an optical medium is quite satisfactorily described by the first two terms of (4.3); that is,

$$P(E) = \chi_1 E + \chi_2 E^2 \tag{4.5}$$

Before proceeding, there is a quite important point to make about (4.3).

Let us consider the effect of a change in sign of E. The two values of the field, $\pm E$, will correspond to two values of P:

$$P(+E) = \chi_1 E + \chi_2 E^2$$

$$P(-E) = -\chi_1 E + \chi_2 E^2$$

These two values clearly have different absolute magnitudes. Now if a medium is isotropic (as is the amorphous silica of which optical fiber is made) there can be no directionality in the medium and thus the matter of the sign of E (i.e., whether the electric field points up or down, cannot be of any physical relevance and cannot possibly have any measurable physical effect). In particular, it cannot possibly affect the value of the electric polarization (which is, of course, readily measurable). We should expect that changing the sign of E will merely change the sign of P, but that the magnitude of P will be exactly the same: the electrons will be displaced by the same amount in the opposite direction, all directions being equivalent. Clearly this can only be so if $\chi_2 = 0$. The same argument extended to higher order terms evidently leads us to the conclusion that all even-order terms *must* be zero for amorphous (isotropic) materials; that is, $\chi_{2m} = 0$. This is a point to remember. The corollary of this argument is, of course, that in order to retain any even-order terms the medium must exhibit some anisotropy. It must, for example, have a crystalline structure without a center of symmetry. It follows that (4.3) refers to such a medium.

Suppose now that we represent the electric field of an optical wave entering such a crystalline medium by

$$E = E_0 \cos \omega t$$

Then substituting into (4.3) we find

$$P(E) = \chi_1 E_0 \cos \omega t + \frac{1}{2}\chi_2 E_0^2 + \frac{1}{2}\chi_2 E_0^2 \cos 2\omega t$$

The last term, the second harmonic term at twice the original frequency, is clearly in evidence. Fundamentally, it is due to the fact that it is easier to polarize the medium in one direction than in the opposite direction, as a result of the crystal asymmetry. A kind of "rectification" occurs.

Now the propagation of the wave through the crystal is the result of adding the original wave to the secondary wavelets from the oscillating dipoles that it induces. These oscillating dipoles are represented by P: thus, $\partial^2 P/\partial t^2$ leads to e/m waves, since radiated power is proportional to the acceleration of charges, and waves at all of P's frequencies will propagate through the crystal.

Suppose now that an attempt is made to generate a second harmonic over a length L of crystal. At each point along the path of the input wave a second harmonic component will be generated. However, since the crystal medium will almost certainly be dispersive, the fundamental and second harmonic components will travel at different velocities. Hence the successive portions of second harmonic component generated by the fundamental will not, in general, be in phase with each other, and thus will not interfere constructively. Hence, the efficiency of the generation will depend upon the velocity difference between the waves.

For maximum efficiency of second harmonic generation, we require the two velocities to be equal, for then all points at which the generation occurs will be in phase. It is shown in Appendix E that the efficiency of the second harmonic generation (SHG) is given by

$$\eta_{SHG} = B\chi_2^2 I_L(\omega) L^2 \omega^2 \left[\frac{\sin\left(k - \frac{1}{2}k_s\right)L}{\left(k - \frac{1}{2}k_s\right)L} \right]^2 \qquad (4.7)$$

where B is a constant. Note that η_{SHG} varies as the square of the fundamental frequency and of the length of the crystal; note also that it increases linearly with the power of the fundamental.

From (4.7) it is clear that, for maximum intensity, we require that the sinc^2 function has its maximum value; that is, that:

$$k_s = 2k_f$$

This is the *phase matching condition* for second harmonic generation. Now the velocities of the fundamental and the second harmonic are given by

$$c_f = \frac{\omega}{k_f}, \; c_s = \frac{2\omega}{k_s}$$

These are equal when $k_s = 2k_f$, so, as expected, the phase matching condition is equivalent to a requirement that the two velocities are equal.

The phase matching condition usually can be satisfied by choosing the optical path to lie in a particular direction within the crystal. It has already been noted that the material must be anisotropic for second harmonic generation to occur; it will also, therefore, exhibit birefringence (Section 3.3). One way of solving the phase-matching problem, therefore, is to arrange that the velocity difference resulting from birefringence is cancelled by that resulting from material dispersion. In a crystal with normal dispersion, the refractive index of both the eigenmodes (i.e., both the ordinary and extraordinary rays) increases with frequency. Suppose we consider the specific example of quartz, which is a positive uniaxial crystal (see Section 3.3). This means that the principal refractive index for the extraordinary ray is greater than that for the ordinary ray; that is,

$$n_e > n_o$$

Since quartz is also normally dispersive, it follows that

$$n_e^{(2\omega)} > n_o^{(\omega)}$$

$$n_o^{(2\omega)} > n_o^{(\omega)}$$

Hence the index ellipsoids for the two frequencies are as shown in Figure 4.10(a). Now it will be remembered from Section 3.3 that the refractive indices for the "o" and "e" rays for any given direction in the crystal are given by the major and minor axes of the ellipse in which the plane normal to the direction, and passing through the center of the index ellipsoid, intersects the surface of the ellipsoid. The geometry [Figure 4.10(a)] thus makes it clear that a direction can be found [6] for which

$$n_o^{(2\omega)}(\vartheta_m) = n_e^{(\omega)}(\vartheta_m)$$

so SHG phase matching occurs provided that

$$n_o^{(2\omega)} < n_e^{(\omega)}$$

The above is indeed true for quartz over the optical range. Simple trigonometry allows ϑ_m to be determined in terms of the principle refractive indices as

$$\sin^2 \vartheta_m = \frac{\left(n_e^{(\omega)}\right)^{-2} - \left(n_e^{(2\omega)}\right)^{-2}}{\left(n_o^{(2\omega)}\right)^{-2} - \left(n_e^{(2\omega)}\right)^{-2}}$$

Hence ϑ_m is the angle at which phase matching occurs. It also follows from this that, for second harmonic generation in this case, the wave at the fundamental frequency must be launched at angle ϑ_m with respect to the crystal axis and *must have the "extraordinary" polarization*; and that the second harmonic component *will appear in the same direction and will have the "ordinary" polarization* (i.e., the two waves are collinear and have orthogonal linear polarization).

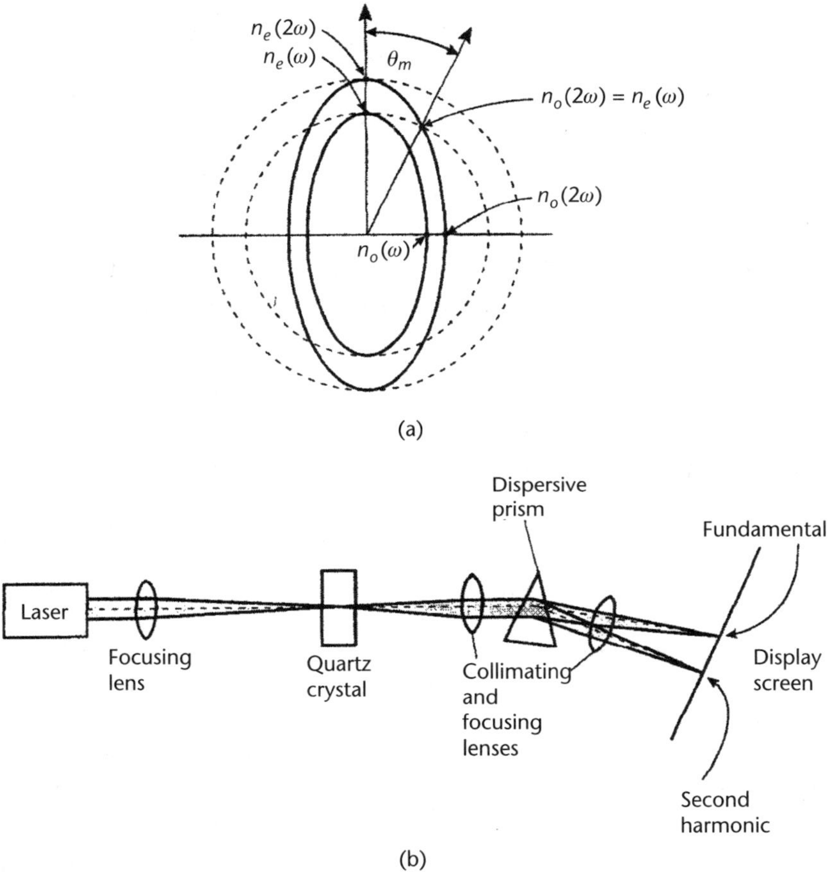

Figure 4.10 Conditions for second harmonic generation in quartz: (a) phase matching with birefringence index ellipsoids; and (b) schematic experimental arrangement for SHM generation.

Clearly, other crystal-direction and polarization arrangements also are possible in other crystals.

The required conditions can be satisfied in many crystals but quartz is an especially good one owing to its physical robustness, its ready obtainability with good optical quality, and its high optical power-handling capacity.

Provided that the input light propagates along the chosen axis, the conversion efficiency ($\omega \rightarrow 2\omega$) is a maximum compared with any other path (per unit length) through the crystal. Care must be taken, however, to minimize the divergence of the beam (so that most of the energy travels in the chosen direction) and to ensure that the temperature remains constant (since the birefringence of the crystal will be temperature dependent).

The particle picture of the second harmonic generation process is viewed as an annihilation of two photons at the fundamental frequency, and the creation of one photon at the second harmonic frequency. This pair of processes is necessary in order to conserve energy:

$$2h\nu_f = h(2\nu_f) = h\nu_s$$

The phase matching condition is then equivalent to conservation of momentum. The momentum of a photon wave number k is given by

$$p = \frac{h}{2\pi} k$$

and thus conservation requires that

$$k_s = 2k_f$$

as in the wave treatment.

Quantum processes which have no need to dispose of excess momentum are again the most probable, and thus, this represents the condition for maximum conversion efficiency in the particle picture.

The primary practical importance of second harmonic generation is that it allows laser light to be produced at the higher frequencies, into the blue and ultraviolet, where conditions are not intrinsically favorable for laser action. In this context we note again, from (4.7), that the efficiency of the generation increases as the square of the fundamental frequency, which is of assistance in producing these higher frequencies.

4.3.5 Optical Mixing

Optical mixing is a process closely related to second harmonic generation. If, instead of propagating just one laser wave through the same nonlinear crystal, we superimpose two (at different optical frequencies) simultaneously along the same direction, then we shall generate sum and difference frequencies:

$$E = E_1 \cos \omega_1 t + E_2 \cos \omega_2 t$$

and thus, again using (9.3),

$$P(E) = \chi_1(E_1 \cos \omega_1 t + E_2 \cos \omega_2 t) + \chi_2(E_1 \cos \omega_1 t + E_2 \cos \omega_2 t)^2$$

This expression for $P(E)$ is seen to contain the term

$$2\chi_2 E_1 E_2 \cos \omega_1 t \cos \omega_2 t = \chi_2 E_1 E_2 \cos(\omega_1 + \omega_2)t + \chi_2 E_1 E_2 \cos(\omega_1 - \omega_2)t$$

giving the required sum and difference frequency terms. Again, for efficient generation of these components, we must ensure that they are phase matched. For example, to generate the sum frequency efficiently we require that

$$k_1 + k_2 = k_{(1+2)}$$

which is equivalent to

$$\omega_1 n_1 + \omega_2 n_2 = (\omega_1 + \omega_2) n_{(1+2)}$$

where n represents the refractive indices at the suffix frequencies. The condition, again, is satisfied by choosing an appropriate direction relative to the crystal axes.

This mixing process is particularly useful in the reverse sense. If a suitable crystal is placed in a Fabry-Perot cavity which possesses a resonance at ω_1, say, and is pumped by laser radiation at $\omega_{(1+2)}$, then the latter generates both ω_1 and ω_2. This process is called parametric oscillation: ω_1 is called the signal frequency and ω_2 the idler frequency. It is a useful method for *down conversion* of an optical frequency (i.e., conversion from a higher to a lower value).

The importance of phase matching in nonlinear optics cannot be overstressed. If waves at frequencies different from the fundamental are to be generated efficiently, they must be produced with the correct relative phase to allow constructive interference, and this, as we have seen, means that velocities must be equal to allow phase matching to occur. This feature dominates the practical application of nonlinear optics.

4.3.6 Intensity-Dependent Refractive Index

It was noted in Section 4.3.4 that all the even-order terms in (4.3) for the nonlinear susceptibility (χ) are zero for an amorphous (i.e., isotropic) medium. This means, of course, that, in an optical fiber, made from amorphous silica, we can expect that $\chi_{(2m)} = 0$, so it will not be possible to generate a second harmonic according to the principles outlined in Section 4.3.4. (However, second harmonic generation has been observed in fibers [7] for reasons that took some time to understand!) It is possible to generate a third harmonic, however, since to a good approximation the electric polarization in the fiber can be expressed by

$$P(E) = \chi_1 E + \chi_3 E^3 \tag{4.8}$$

Clearly, though, if we wish to generate the third harmonic efficiently we must again phase match it with the fundamental, and this means that somehow we must arrange for the two relevant velocities to be equal; that is, $c_\omega = c_{3\omega}$. This is very difficult to achieve in practice, although it has been done.

There is, however, a more important application of (4.8) in amorphous media. From Section 4.2 we know that the effective refractive index in this case can be written:

$$n_e = \left(1 + \chi_1 + \chi_3 E^2\right)^{1/2}$$

and, if $\chi_1, \chi_3 E^2 \ll 1$,

$$n_e \approx 1 + \frac{1}{2}\chi_1 + \frac{1}{2}\chi_3 E^2$$

Hence,

$$n_e = n_o + \frac{1}{2}\chi_3 E^2 \tag{4.9a}$$

where n_o is the "normal," linear refractive index of the medium. But we know that the intensity (power/unit area) of the light is proportional to E^2, so that we can write

$$n_e = n_o + n_2 I \tag{4.9b}$$

where n_2 is a constant for the medium. Equation (4.9b) is very important and has a number of practical consequences. We can see immediately that it means that the refractive index of the medium depends upon the intensity of the propagating light: the light is influencing its own velocity as it travels.

In order to fix ideas to some extent, let us consider some numbers for silica. For amorphous silica $n_2 \sim 3.2 \times 10^{-20}$ $\mathrm{m^2W^{-1}}$, which means that a 1% change in refractive index (readily observable) will occur for an intensity $\sim 5 \times 10^{17}$ $\mathrm{Wm^{-2}}$. For a fiber with a core diameter ~ 5 μm this requires an optical power level of 10 MW. Peak power levels of this magnitude are readily obtainable, for short durations, with modern lasers.

It is interesting to note that this phenomenon is another aspect of the electro-optic effect that was described in Section 4.2.6. Clearly the refractive index of the medium is being altered by an electric field. This will now be considered in more detail.

4.3.7 Optical Kerr Effect

The normal electro-optic Kerr effect was considered in Section 4.2.6. It is an effect whereby an electric field imposed on a medium induces a linear birefringence with slow axis parallel with the field [Figure 4.11(a)]. The value of the induced birefringence is proportional to the square of the electric field.

In the optical Kerr effect the electric field involved is that of an optical wave, and thus the birefringence probed by one wave may be that produced by another. In Figure 4.11(b) a forward-propagating wave, linearly polarized at 45° to the vertical, meets a counterpropagating, high-intensity pulse, linearly polarized in the vertical direction. This pulse induces a phase change, via the Kerr effect of its electric field, in the vertical component of the forward-propagating wave, thus altering its polarization state. Hence we have an effect of "light acting upon light," one of the applications of which is very fast switching: the polarization state change occurs over the duration of what can be a very short counter-propagating pulse.

The phase difference introduced by an electric field E over an optical path L is given by

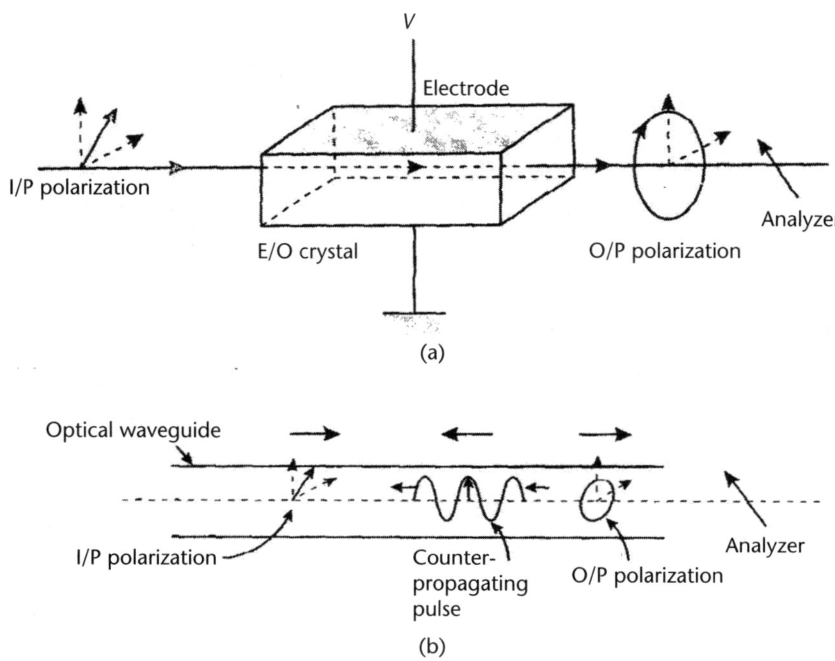

Figure 4.11 (a) "Normal" electro-optic Kerr effect. (b) "Optical" Kerr effect: light acting on light.

$$\Delta\varphi = \frac{2\pi}{\lambda}\Delta n \cdot L$$

where $\Delta n = KE^2$, K being the Kerr constant.

Now from (4.9a) and (4.9b) we have

$$\Delta n = n_2 I = \frac{1}{2}\chi_3 E^2 = KE^2 \tag{4.10}$$

From our discussions of elementary electromagnetism in Section 1.2.2 we derived the equation:

$$I = c\epsilon E^2$$

Hence we have, from (4.10)

$$K = n_2 c\epsilon = \frac{1}{2}\chi_3$$

showing that the electro-optic effect, whether the result of an optical or an external electric field, is a nonlinear phenomenon, depending on χ_3. Using similar arguments, it can easily be shown that the electro-optic Pockels effect (Section 4.2.6) also is a nonlinear effect, depending on χ_2. (Remember that the Pockels effect can only occur in anisotropic media, so that χ_2 will be nonzero.)

The optical Kerr effect has several other interesting consequences. One of these is self-phase modulation, which is the next topic for consideration.

4.3.8 Self-Phase Modulation

The fact that refractive index can be dependent on optical intensity clearly has implications for the phase of the wave propagating in nonlinear medium. We have

$$\varphi = \frac{2\pi}{\lambda} nL$$

Hence, for $n = n_o + n_2 I$,

$$\varphi = \frac{2\pi L}{\lambda}(n_o + n_2 I)$$

Suppose now that the intensity is a time-dependent function $I(t)$. It follows that φ also will be time dependent, and, since

$$\omega = \frac{d\varphi}{dt}$$

the frequency spectrum will be changed by this effect, which is known as *self-phase modulation* (SPM).

In a dispersive medium a change in the spectrum of a temporally varying function (e.g., a pulse) will change the shape of the function. For example, pulse broadening or *pulse compression* can be obtained under appropriate circumstances. To see this, consider a Gaussian pulse [Figure 4.12(a)]. The Gaussian shape modulates an optical carrier of frequency ω_0, say, and the new instantaneous frequency becomes

$$\omega' = \omega_0 + \frac{d\varphi}{dt}$$

If the pulse is propagating in the *Oz* direction,

$$\varphi = -\frac{2\pi z}{\lambda}(n_o + n_2 I) \tag{4.11a}$$

and we have

$$\omega' = \omega_0 - \frac{2\pi z}{\lambda} n_2 \frac{dI}{dt} \tag{4.11b}$$

At the leading edge of the pulse, $dI/dt > 0$, hence

$$\omega' = \omega - \omega_I(t)$$

At the trailing edge

$$\frac{dI}{dt} < 0$$

and hence

$$\omega' = \omega + \omega_I(t)$$

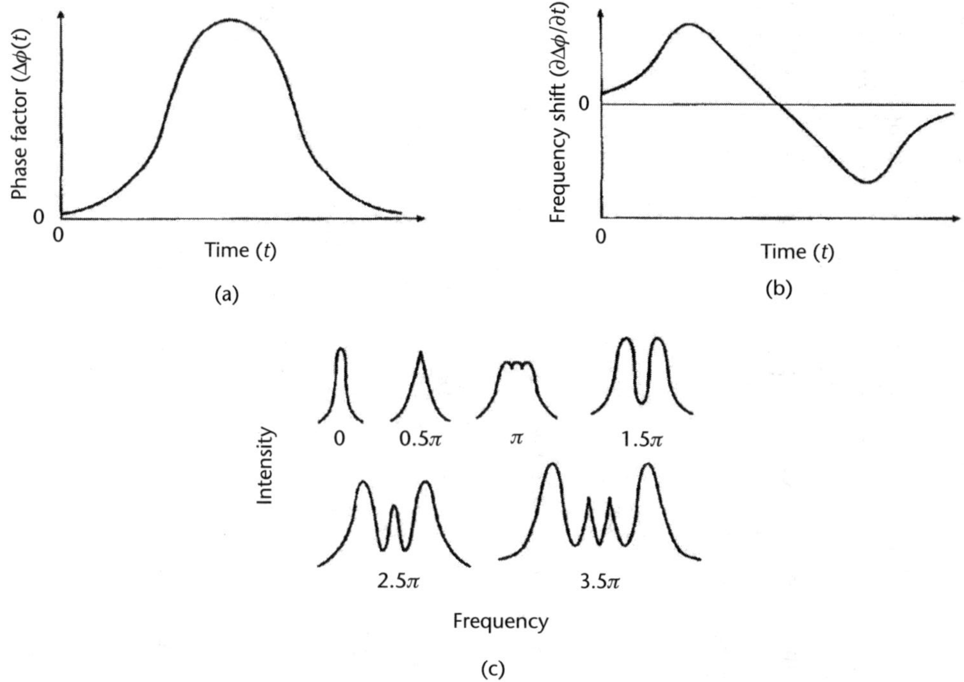

Figure 4.12 Self-phase modulation for a Gaussian pulse: (a) intensity-dependent phase factor for a Gaussian pulse; (b) the instantaneous frequency shift for (a); and (c) frequency spectra for (a) designated by maximum phase shift, at peak. (*From:* [10]. © 1978 *Phys. Rev. A*. Reprinted with permission.)

The pulse is now *chirped* (i.e., the frequency varies across the pulse). Figure 4.12(b) shows an example of this effect.

Suppose, for example, a pulse from a mode-locked Argon laser, with an initial width of 180 ps, is passed down 100m of optical fiber. As a result of self-phase modulation, the frequency spectrum is changed by the propagation. Figure 4.12(c) shows how the spectrum varies as the initial peak power of the pulse is varied. The peak power will lead to a peak phase change, according to (4.10a), and this phase change is shown for each of the spectra. It can be seen that the initial spectrum ($\Delta\varphi = 0$) is due just to the modulation of the optical sinusoid (Fourier spectrum of a Gaussian pulse) and, as the value of $\Delta\varphi$ increases, the first effect is a broadening.

At $\Delta\varphi = 1.5\pi$ the spectrum has split into two clear peaks, corresponding to the frequency shifts at the back and front edges of the pulse. The spectra then develop multiple peaks.

It is important to realize that this does not necessarily change the shape of the pulse envelope, just the optical frequency within it. However, if the medium through which the pulse is passing is dispersive, the pulse shape will change. This is an interesting possibility and it will be considered further in Section 4.4.

4.3.9 Four-Wave Mixing (FWM)

In Sections 4.3.3 and 4.3.4 we saw how two photons of certain frequencies could be "mixed" to generate photons at different frequencies in the processes of second harmonic generation, and of sum and difference frequency generation. These processes were "mediated" (as we say) by χ_2. In Section 4.3.7 we saw how the electro-optic effect in amorphous media (Kerr effect) was mediated by χ_3. Can χ_3 be used to generate new frequencies? If it can, it would be very convenient if it could do so in amorphous silica because then it could be used with the high-intensity, long path lengths associated with optical fibers, and efficient generations could be expected. The two-photon mixing processes considered in the preceding sections relied upon the mixing of two fields, and thus on two photons, in the squared-field second term of (4.3): $\chi_2 E^2$. If χ_3 is to be used via the third term, $\chi_3 E^3$, we naturally expect, therefore, that three photons will be involved.

Let us consider three optical waves of frequencies ω_p, ω_s, and ω_a and further suppose (for reasons which will soon become clear) that these are related by

$$2\omega_p = \omega_s + \omega_a \qquad (4.12)$$

(Note that four photons are involved here).

It is clear that, under this condition, the term $\chi_3 E^2_{\omega_p} E_{\omega_a}$ will generate a frequency:

$$2\omega_p - \omega_a = \omega_s$$

and that the term $\chi_3 E^2_{\omega_p} E_{\omega_s}$ will generate a frequency:

$$2\omega_p - \omega_s = \omega_a$$

Hence, ω_s and ω_a are continuously generated by each other, with the assistance of ω_p and χ_3. However, as we know very well now, this can only take place efficiently if there is phase matching (i.e., if the velocities of ω_p, ω_s, and ω_a are all the same).

Interestingly, this can be achieved using high linear birefringence (hi-bi; Section 4.2.1) fiber. Remember that the two linearly polarized eigenmodes in such a fiber have different velocities. Also remember that there is material and waveguide dispersion to take into account. The result is that, provided that the "pump" frequency (ω_p) is chosen correctly in relation to the dispersion characteristic and launched as a linearly polarized wave with its polarization direction aligned with the fiber's slow axis, then, as in the case of second harmonic generation in a crystal, the combination of the velocity difference in the other polarization eigenmode (fast axis) and the dispersion in the fiber (material and waveguide) can allow two other frequencies, ω_s and ω_a, to have the same velocity in the fast mode as ω_p has in the slow mode, and also to satisfy (4.12). ω_s is called the Stokes frequency and ω_a the anti-Stokes frequency when $\omega_a > \omega_s$. This process clearly involves four waves (ω_p, ω_p, ω_a, ω_s) and hence the name *four-wave mixing* (FWM).

The process is analogous to that known as parametric downconversion in microwaves, where it is used to produce a downconverted frequency (ω_s) known as the "signal" and an (unwanted) upconverted frequency (ω_a) known as the "idler." An optical four-wave mixing frequency spectrum generated in hi-bi fiber is shown in Figure 4.13.

Four-wave mixing has a number of uses. An especially valuable one is that of an optical amplifier. If a pump is injected at ω_p, it will provide gain for signals injected (in the orthogonal polarization of course) at ω_s or ω_a. The gain can be controlled by injecting signals at ω_s and ω_a simultaneously, and then varying their relative phase. The pump will provide more gain to the component which is the more closely phase matched (Figure 4.14).

Another useful application is that of determining which is the "fast" and which the "slow" axis of a hi-bi fiber. Only when the pump is injected into

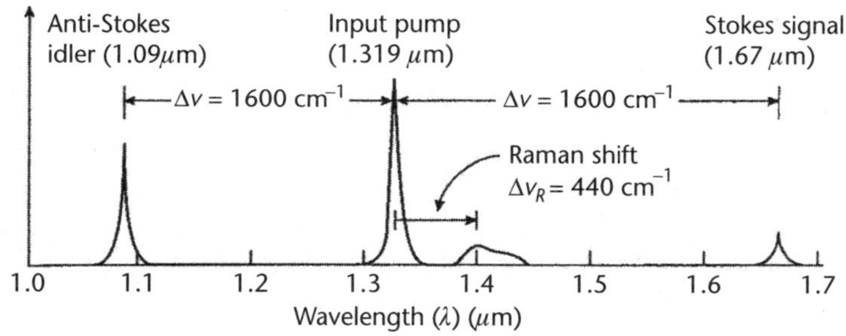

Figure 4.13 Four-photon mixing spectrum in hi-bi fiber. (*From:* [11]. © 1981 *Opt. Lett.* Reprinted with permission.)

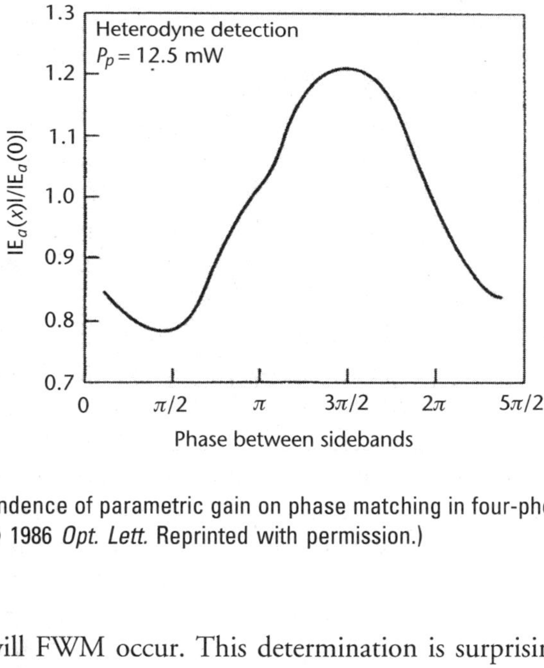

Figure 4.14 Dependence of parametric gain on phase matching in four-photon mixing. (*From:* [8]. © 1986 *Opt. Lett.* Reprinted with permission.)

the slow axis will FWM occur. This determination is surprisingly difficult by any other method. By measuring accurately the frequencies ω_a and ω_s, variations in birefringence can be tracked, implying possibilities for use in optical-fiber sensing of any external influences which affect the birefringence (e.g., temperature, stress).

Finally, the effects of FWM can be unwanted also. In optical-fiber telecommunications the generation of frequencies other than that of the input signal, via capricious birefringence effects, can lead to cross-talk in multichannel systems—for example, wavelength division multiplexed (WDM) systems—and hence must be avoided.

4.4 Solitons

No account of nonlinear optics in optical fibers, even at this physical level, would be complete without a brief discussion of optical solitons. A soliton is a solitary wave, a wave pulse which propagates, even in the face of group velocity dispersion (GVD) in the medium, over long distances without change of form. It thus has enormous potential for application to long-distance optical-fiber digital communications.

The soliton is not limited to optics. It comprises a particular set of solutions of the nonlinear Schrodinger wave equation and is a possibility whenever nonlinear wave motion occurs in a dispersive medium. It was first observed in 1834 (before any theory was worked out and certainly long before Schrodinger formulated his wave equation) as a large amplitude water wave propagating along the Union Canal which connects Edinburgh and Glasgow, in Scotland. John Scott Russell, a Scottish civil engineer, was exercising his horse alongside the canal in the summer of 1834 when he noticed a wave pulse which had been generated by the sudden halting of a horse-drawn barge. This wave traveled without change of shape or size, and Russell followed it on his horse for two kilometers, noting how puzzlingly stable it was. This turned out to be a water-wave soliton, but solitons were subsequently observed in many other branches of physics, wherever wave motion can occur, in fact, for all restorative systems are capable of being drive into nonlinearity.

The detailed mathematical analysis of this phenomenon is complex, but the basic ideas are relatively straightforward; moreover, they follow from ideas that we have already discussed. In Section 4.3 we studied group velocity dispersion, and earlier in this chapter (Section 4.3.8) we studied self-phase modulation. These two ideas must be put together in order to understand solitons.

Let us, as in Section 4.3.7, take an optical wave pulse with a Gaussian intensity envelope, and pass it into a dispersive medium. Since the source of the pulse will have a nonzero spectral width, the GVD will act on this pulse to broaden it: the positive GVD will cause the lower frequencies to arrive at the output first, and the negative GVD the higher frequencies first. Clearly, in both cases the pulse is broadened [Figure 4.15(a)]. If SPM also is present, however, we know that, from the consequences of (4.11b), higher frequencies are produced at the trailing edge of the pulse, while lower ones are produced at the leading edge [Figure 4.15(b)]. If this effect happens in the presence of negative GVD, then the trailing edge of the pulse will tend to catch up with leading edge: the pulse will be compressed [Figure 4.15(c)]. It is easy to see that under certain circumstances this compressive effect might exactly balance the spreading effect due to the source spectral width, and the pulse width can then remain constant throughout the propagation.

This balance can indeed be struck and the result is a soliton.

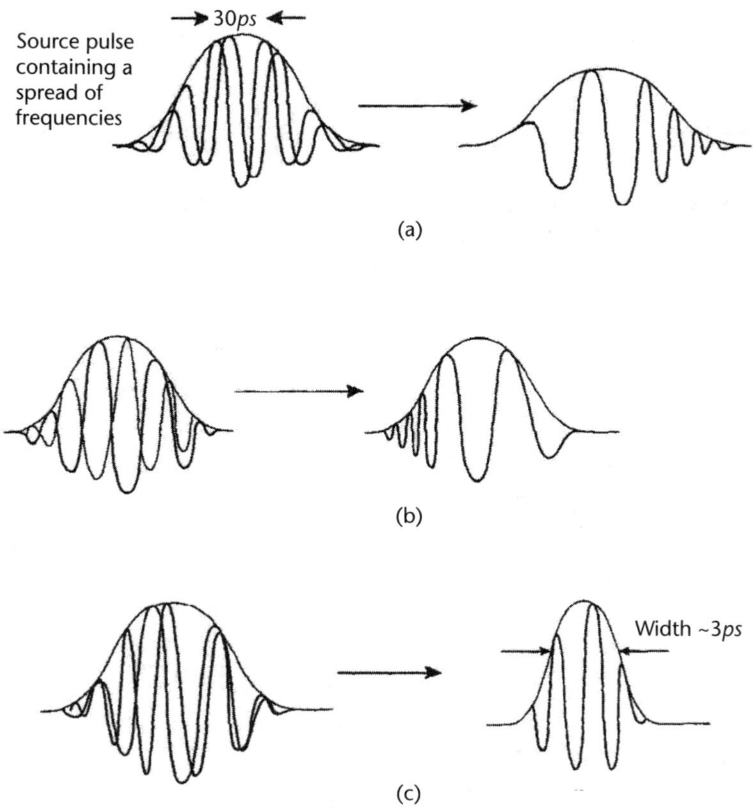

Figure 4.15 Essentials of soliton formation: (a) negative GVD–high frequencies have greater velocity than low frequencies and the pulse is broadened; (b) self-phase modulation—high frequencies are chirped to the trailing edge of the pulse; and (c) self-phase modulation together with negative GVD—the effects in (a) and (b) balance to a stable soliton.

Solitons have been observed in optical fibers (Figure 4.16) and could lead to optical communications systems of phenomenal bandwidth × distance products, perhaps as high as 10,000 GHz km. However, the theory shows solitons to be unstable in lossy media. In addition they tend to attract each other when closer than ~ 10 pulse widths apart. These features clearly limit their advantages in the communications area.

Their potential remains considerable, however, and the research into them undoubtedly will continue.

An interesting corollary to the discussion of the effects which give rise to solitons involves the interaction of a Gaussian pulse with a positive GVD. Figure 4.17 shows the effect, on the spectrum of this pulse, of SPM with a *positive* GVD medium. It is seen there that the result is a much extended region of

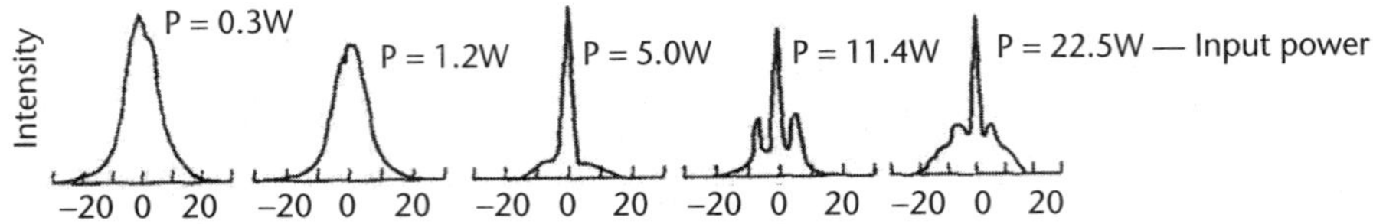

Figure 4.16 Measured solitons emerging from an optical fiber. (*From:* [9]. © 1980 *Phys. Rev. Lett.* Reprinted with permission.)

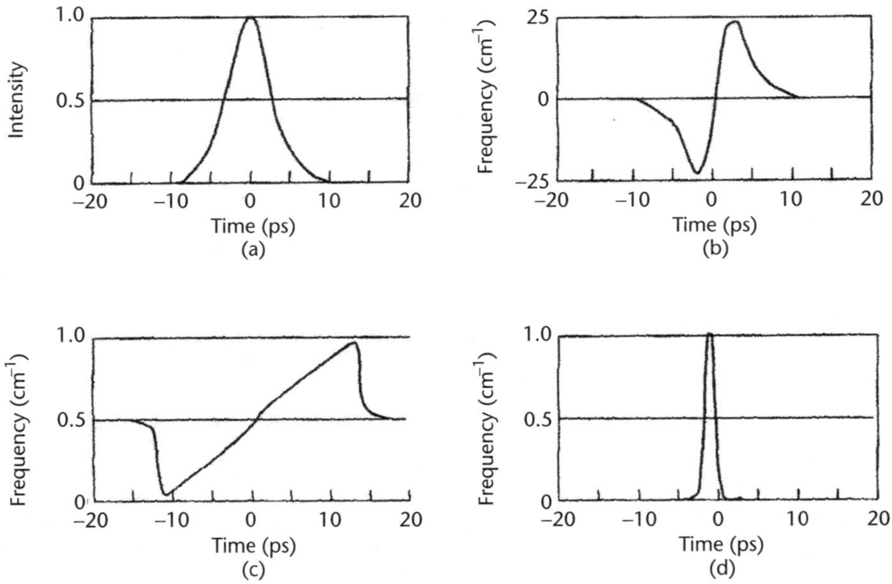

Figure 4.17 Pulse compression using SPM with positive GVD followed by negative GVD: (a) narrow input pulse profile; (b) pulse spectrum after SPM; (c) pulse spectrum with positive GVD; and (d) compressed pulse after negative GVD acting on spectrum from (c).

linear chirp [Figure 4.17(c)] where the frequency varies linearly from the front edge to the back edge of the pulse. If, after this has been done, the pulse is then passed into a purely *negative* GVD medium (e.g., another fiber), the pulse can be very strongly compressed. Pulse widths as small as 8 fs (8×10^{-15} s) have been produced, using such a method, at a wavelength of 620 nm. Such pulses contain only about four optical cycles!

Pulses such as these can be used in research to study, for example, very fast molecular processes, single-atom chemical reactions, and very fast switching phenomena.

4.5 Conclusions

This chapter has dealt with, in a fairly general, physical way, a range of polarization effects in optical fibers. We have seen that these effects fall quite naturally into two groupings: the linear and nonlinear processes.

The linear processes are largely concerned with form and asymmetry in the fiber waveguide, either intrinsic or imposed externally. The nonlinear processes result from a modulation of the material properties by the propagating light itself, and rely heavily on the concept of phase matching.

In the next two chapters we shall learn how better to quantify and to control these effects, either for the purposes of mitigation where they are an unwanted source of interference or in order to use them positively to make quantitative measurements of wanted parameters.

References

[1] Barlow, A. J., et al., "Production of Single-Mode Fibers with Negligible Intrinsic Birefringence and Polarization Mode Dispersion," *Elect. Lett.*, Vol. 17, 1981, pp. 725–726.

[2] Varnham, M. P., et al., "Single Polarization Operation of Highly Birefringent Bow-Tie Optical Filters," *Elect. Lett.*, Vol. 19, 1983, pp. 246–247.

[3] Dyott, R. B., *Elliptical Fiber Waveguides*, Norwood, MA: Artech House, 1995.

[4] Ulrich, R., et al., "Bending-Induced Birefringence in Single-Mode Fibers," *Opt. Lett.*, Vol. 5, No. 6, 1980, pp. 273–275.

[5] Ulrich, R., and A. Simon, "Polarization Optics of Twisted Single-Mode Fibers," *Appl. Opt.*, Vol. 18, No. 13, 1979, pp. 2241–2251.

[6] Nye, J. F., *Physical Properties of Crystals*, Oxford, U.K.: Clarendon Press, 1976, Chapter 13.

[7] Fujii, Y., et al., "Sum-Frequency Light Generation in Optical Fibers," *Opt. Lett.*, Vol. 5, 1980, p. 48.

[8] Bar-Joseph, I., et al.. "Parametric Interaction of a Modulated Wave in an SM Fiber," *Opt. Lett.*, Vol. 11, 1986, p. 534.

[9] Mollenauer, L. F., R. H. Stolen, and J. P. Gordon, "Experimental Observation of Picosecond Narrowing and Solitons in Optical Fibers," *Phys. Rev. Lett.*, Vol. 45, 1980, p. 1095.

[10] Stolen, R. H., and C. Jin, "Self-Phase Modulation in Optical Fibres," *Phys. Rev. A*, Vol. 17, 1978, pp. 1448–1452.

[11] Jin, C., et al., "Phase Matching in the Minimum Chromatic Dispension Region of SM Fibres for Stimulated FPM," *Opt. Lett.*, Vol. 6, 1981, pp. 493–495.

5

Practical Applications of Polarization Effects in Optical-Fiber Communications

5.1 Introduction

In the preceding four chapters the groundwork has been laid for an appreciation of the practical ways in which polarization effects in fibers make their presence felt, sometimes deleteriously, and the practical uses to which they can be put. It is to these ideas that we now turn our attentions.

The two primary application areas for optical fibers are those of telecommunications and measurement sensing. We shall deal with each of these, in turn, in the next two chapters.

5.2 Optical Communications Systems

The basic features relevant to the use of optical fibers in communications were described in Section 2.6, and Figure 5.1 shows, again, the elements of a telecommunications system.

The light source, typically, is a semiconductor laser, and its light output power is modulated by the information to be transmitted, which usually is in a digital format (i.e., the light output consists of a stream of pulses comprising the 1s and 0s of a digital bit-stream). The integrity of the communications channel clearly depends upon the ability of the detection system to identify the individual pulses (or absence of pulses) independently of each other. This must

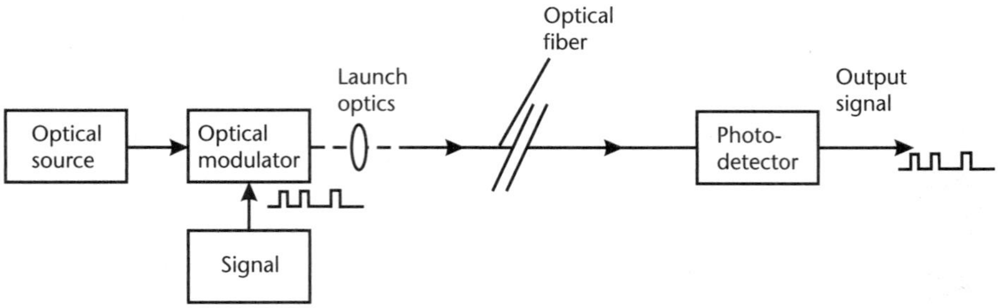

Figure 5.1 Schematic for an optical fiber communications system.

be done in the face of the tendency of the pulses to broaden as they propagate down the fiber. This broadening can be due to several physical effects operating in the fiber, three of which were described in Section 2.6.2: modal dispersion, material dispersion, and waveguide dispersion. None of these is significantly polarization-dependent in normal fibers. However, there is yet another source of fiber dispersion which certainly is so dependent: this is polarization-mode dispersion (PMD) and this phenomenon is assuming increasing importance as the other effects have been successively (and largely successfully) mitigated, and as the pressure towards larger and larger bit-rates, for communications channels, has increased.

There are also some well-used components and devices in the fiber communications systems which are concerned with polarization effects in fibers, and we shall begin by looking at these.

5.3 Polarization Phenomena in Components and Devices for Optical Communications

In general, we require that optical communications systems are polarization independent, especially for the detection process. Exceptions, however, are when polarization modulation is used, or when the system utilizes the advantages of phase coherence. We shall return to these two topics shortly but, for the most part, we might assume that the detection devices for the system are polarization independent and that, therefore, any changes in the polarization state of the light produced by the fiber will not affect link performance. Unfortunately, things are not this straightforward, as a result of three primary phenomena: polarization-dependent loss (PDL), polarization-dependent gain (PDG), and polarization mode dispersion (PMD). We begin by looking at PDL and PDG, followed by some polarization components. We shall discuss PMD at length in Section 5.4.

Polarization-Dependent Loss

PDL is the result of losses which vary with the polarization state as a result of transverse directionality in the loss mechanism. Normally, the PDL present in an optical fiber used for communications will be very small. The loss on bending a fiber will be polarization dependent because the plane of the bend will introduce the directionality, causing the linear state in the plane of the bend to suffer more loss than the orthogonal direction, but bend radii will be quite large for any installed system and the cable design will, in any case, resist any sharp bend. Most of the PDL will result from system components such as splitters, couplers, filters, switches—all of which will inevitably possess some directionality but will be designed to minimize it.

PDL is quantified by a measure of the ratio (expressed in decibels) of the maximum transmitted power to the minimum transmitted power, when viewed over all possible polarization states. It is normally measured by viewing the power spectrum for a random scan over polarization states, using a polarization controller (see below). It may be defined formally by

$$PDL = 10 \log_{10}(T_{\max}/T_{\min}) \tag{5.1}$$

where $T_{\max}$ and $T_{\min}$ are the maximum and minimum transmitted powers when viewed over all polarization states.

Polarization-Dependent Gain

PDG results largely from the use of optical-fiber amplifiers. In very long-haul systems (> 10,000 km, say) several hundred of these may be used, so that the effect can be significant.

An optical-fiber amplifier operates via pump light which excites a dopant atom, usually erbium, into an excited state. This excited state then either decays back to the ground state spontaneously or is stimulated to do so by the incoming light signal. In both cases the photon so produced continues on to de-excite other states.

If the de-excitation is spontaneous, the resulting amplified signal leads to an output known as amplified spontaneous emission (ASE) and comprises a noise signal. Clearly, the amplification resulting from the signal input will lead to the wanted amplified signal.

Now the input signal will quickly de-excite those excited atoms whose dipoles are aligned with its own polarization state. This means that, at the orthogonal polarization, the ASE is free to enjoy all the excitations, so that it is enhanced, thus increasing the noise. This phenomenon is called polarization-hole burning (PHB), and leads to a reduction in the SNR at the receiver. The effect can be reduced to low levels by introducing some birefringence into the amplifying fiber. In this case the signal polarization (not, in general, aligned with an eigenmode) will sweep through a range of states, thus averaging to zero the PHB.

The PDG effect is quantified formally just as for PDL.

Polarization Controllers

It is often necessary, in telecommunications and in many other areas of application, to have control over the polarization state of the light. We saw in Section 3.4 the way in which a linear polarization can be converted into an elliptical state, with any ellipticity and orientation, using just two waveplates, a $\lambda/2$ and a $\lambda/4$.

We now consider the more general problem of converting any input state into any output state, and by means of fiber-based components. A manual method for achieving this is shown in Figure 5.2. A fiber is wound around two spools so that the bend birefringence in the first endows it with a linear birefringence retardation (at the wavelength of interest) of $\pi/2$, whilst that in the second has one of π (i.e., a $\lambda/4$ retarder followed by a $\lambda/2$).

The plane of each of the spools is rotatable about the fiber axis so that the "plates" can have any orientation with respect to the incoming polarization state.

The action of this controller is best viewed as rotations on the Poincaré sphere, projected onto suitable planes, for simplicity. Here we consider the plane within which occurs the rotation due to the first retarder.

Consider the input state of polarization (SOP) at point A (Figure 5.3) and suppose it is to be converted into the SOP at point C. We require the first rotation to provide a state with the required ellipticity and this can be achieved by determining the plane of the $\pi/2$ rotation which takes A to that latitude on the sphere. If the eigen-axes of the first spool are then orientated normal to this plane (shown in Figure 5.3), the required rotation will occur to point B. It is then just a matter of orienting the second spool, essentially a rotator, so

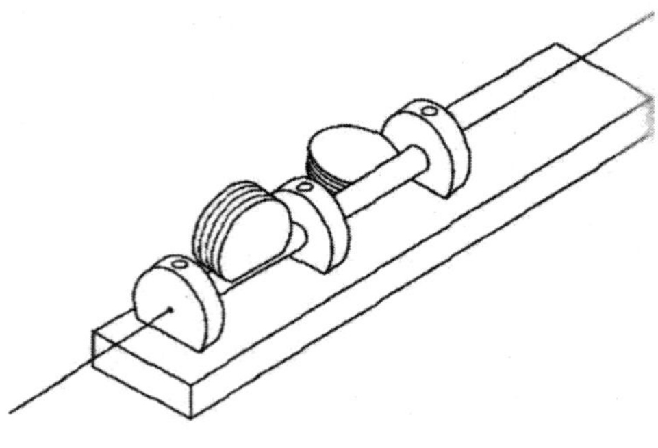

Figure 5.2 Two-stage polarization controller.

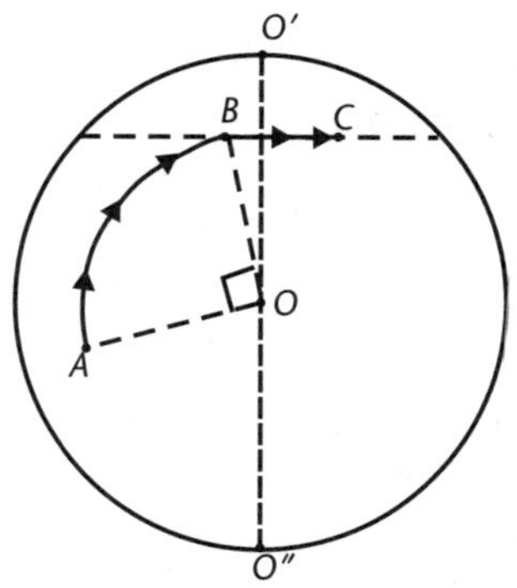

Figure 5.3 Action of a two-stage polarization controller.

that the rotation occurs about the polar axis to bring the elliptical state to the correct azimuthal orientation at the point C.

Quite often, however, the controlling process needs to be automated. (An important case in point is the mitigation of PMD, to be considered in Section 5.4.)

For the automated process to take place within a single fiber, a first requirement is for no moving elements. A rotation could, conceivably, be effected via a Faraday rotation but the Faraday effect is very small in silica fibers and very long, doped lengths of fiber would be required, together with large solenoids to provide the magnetic fields. Such arrangements are impracticable.

One solution is shown in Figure 5.4. Here we have three sections of fiber, each of which is subjected to pressure which induces some linear birefringence, either magneto-strictively or piezo-electrically, at fixed angles of 0°, 45°, 0°, respectively. Each section of fiber thus imposes its retardation under continuous control over the full range of 0 to 2π.

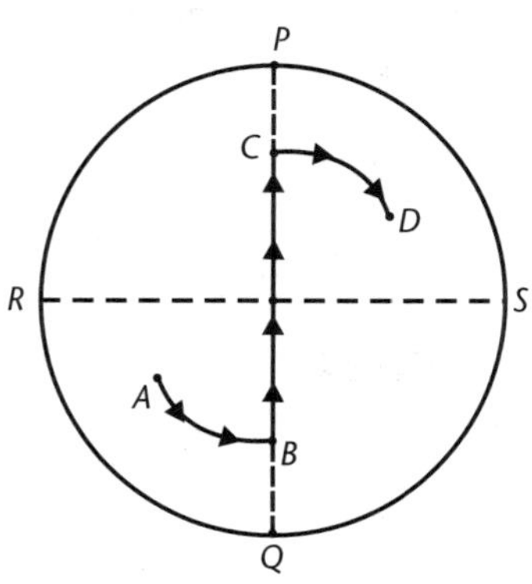

Figure 5.4 Action of a three-stage polarization controller.

Looking, again, to the actions on the Poincaré sphere it is clear that, since we are again dealing with pure retarders, the two eigenaxes, PQ, RS, will lie in the equatorial plane (Figure 5.4) and will be orthogonal to each other.

Consider now an input state A and an output state D. If the final rotation, about PQ, is to rotate on to D, then D and the penultimate state C must lie on a circle whose plane is normal to PQ. For the second rotation to end on the circle containing D and C, the point B could lie on the great circle normal to the plane containing D and C. The action of the first rotation is to take the input state on to that great circle (see Figure 5.4).

Hence it is clear that suitable electrical signals applied to the "squeezers" can convert between any two states with this arrangement. This, therefore, can form the basis for another type of polarization modulator.

A Fiber Polarizer

One method for polarizing light whilst it is propagating within a fiber is shown in Figure 5.5. One side of the cladding of the fiber is polished flat (now a "D" fiber), and a thin metallic film is coated onto the flat surface. The component of the light which is linearly polarized parallel to the plane of the film will suffer relatively small loss because, at the interface, the boundary conditions require that the electric field, E, is zero parallel to the surface of a (perfect) conductor, so that the electric field generated is in anti-phase with the input component, and the light is thus simply reflected. The polarization component of the light which is orthogonal to the film, however, must respond to the condition that D is continuous, and this can only occur if

$$\epsilon_g E_i = \epsilon_m E_r$$

where ϵ_g is the dielectric constant of the guide and ϵ_m that of the metal, so that E_r, the refracted component of the wave's electric field in the metal, is nonzero. The net result is that this field component acts upon the free electrons in the metal, causing them to oscillate against the resistance of the fixed ions.

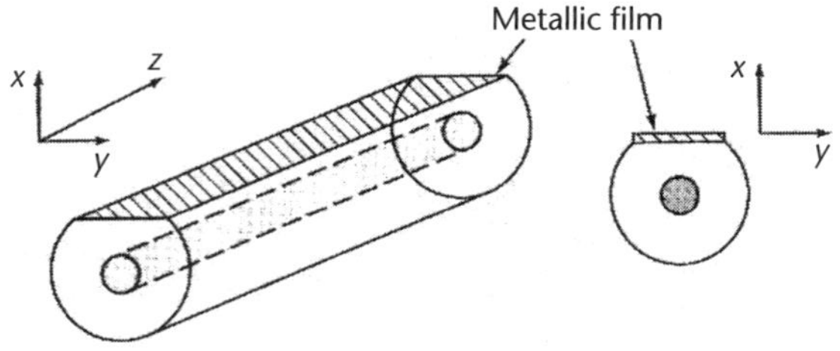

Figure 5.5 A metal-film fiber polarizer.

The oscillation is quantized and is known as a surface plasmon. This oscillation is very lossy, leading to severe loss for this polarization component of the guided wave in the fiber. The consequence is that, after a short distance (~1 cm) only the polarization component parallel to the film persists, and thus the light is effectively linearly polarized.

A full analysis would include all the geometrical parameters in addition to the particular modes which were propagating at a given wavelength; this is quite complex and will not be detailed here, but the physical concept is clear.

5.4 Polarization-Mode Dispersion (PMD)

We saw in Section 2.4 that any deviation of the fiber structure from a uniform material with true cylindrical form, owing to transverse stresses on the optical fiber and/or some loss of exact circularity in the core cross-section, will lead to a difference in phase and group velocities for the two polarization eigenmodes of the fiber. The polarization form of the eigenmodes themselves also will depend on the nature of the perturbations imposed on the ideal form of the fiber.

Fibers can be deliberately stressed or shaped (as was noted in Section 4.2.2) so as to exhibit *high birefringence* (i.e., a large difference in refractive index between the eigenmodes), and such fibers have special uses, some of which we shall discuss in the next chapter. However, all fibers differ to some extent from the ideal. Manufacturing processes, even as sophisticated and refined as they have become, always will produce fibers with some measure of noncircularity and some degree of residual stress.

In the manufacture of high-bandwidth telecommunications fibers, the requirement, firstly, is for monomode (as opposed to multimode) fibers and, secondly, for fiber structures which are as close to the ideal, stress-free, truly cylindrical form as possible.

The reasons for these requirements are, firstly, to remove modal dispersion (monomode fiber) and, secondly, to ensure that there is little dependence of group velocity on polarization state, for such dependence will introduce yet another dispersion effect which will limit the available bandwidth: any given pulse will spread as a result of the differences in arrival time consequent upon the two eigenmode velocities. The result of this, as we shall see in more detail shortly, is to produce fibers which manifest a pulse spreading which increases as the square root of the fiber length, and is of the order (in the best presently available fibers) 0.1 $ps.km^{-1/2}$.

With in-line optical amplifiers—usually erbium-doped fiber amplifiers (EDFAs)—it is possible to install optical-fiber communication systems with effective path lengths in excess of 10,000 km. If we allow pulse spreading only up to 10% of the bit period, the implication is that PMD becomes a problem for bit-rates in excess of 10 Gbs^{-1} over such distances, bandwidths which are increasingly required. Clearly, the greater is the spreading of the pulse in relation to the time between the bits (the bit period), the more difficult it becomes to distinguish between the pulses, and hence the greater is the error rate for the communications system, and the greater the degradation of the system performance.

It is therefore necessary to study PMD in some detail, not least in order to attempt to mitigate it, in pursuit of higher bit-rate communication systems.

5.4.1 Dependence on Optical Path Length

5.4.1.1 "Short" Fibers

We noted in Section 3.8 that any uniform polarization element possesses two polarization eigenmodes, elliptical states in general, which propagate at different velocities through the element. Any polarization state input to the element can be resolved into components with these polarizations, and recombined at the output, including the relative phase delay appropriate to the velocity difference, to provide the output state. The mathematical manipulation for this process is simply the operation of the Jones matrix on the input state, the eigenmodes being those of the matrix itself.

We saw in Section 4.2.2 that we may define a beat length by

$$b = \lambda / \Delta n$$

where Δn is the refractive index difference for the two eigenmodes and λ is the wavelength of the propagating light. Let us suppose that the input is in the form of a pulse of light, equally split between the two eigenmodes. For the special case of linear eigenmodes (see Figure 5.6) the result is that the input pulse splits into a pair of pulses, one for each eigenmode, with a delay, τ,

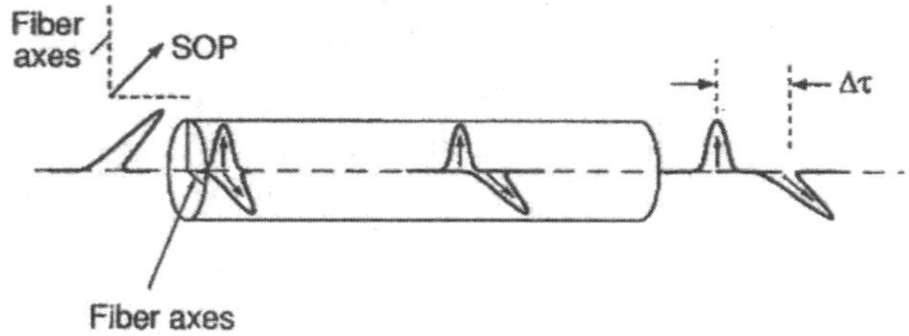

Figure 5.6 Separation of eigenmodes in a short fiber. (*From:* [1]. © 1997 Academic Press. Reprinted with permission.)

between them. If the refractive index of the fiber were independent of wavelength, we could write:

$$\tau = L(n_1 - n_2)/c_0 = L\Delta n/c_0$$

where L is the length of the fiber, c_0 the velocity of light in free space, and n_1, n_2 the refractive indices for the two eigenmodes. However, in practice n_1 and n_2 do depend upon the wavelength, so it is the group velocity difference (see Section 2.6.2), which is active for the pulse splitting and hence, from the equation

$$c_g = d\omega/dk$$

we have

$$\tau = L(1/c_{g1} - 1/c_{g2}) = L\frac{d}{d\omega}(k_1 - k_2)$$

Now

$$k/\omega = n/c_0$$

hence,

$$\tau = L/c_0 \cdot \frac{d}{d\omega}(\omega \cdot \Delta n) = L\Delta n/c_0 + L\omega/c_0 \cdot \frac{d}{d\omega}(\Delta n) \qquad (5.2a)$$

τ is the defining parameter for PMD and, in regard to a uniform Section of fiber, it is clear that it will increase linearly with the length, L, of fiber. τ is thus the delay caused by the group velocity difference, and is known as the differential group delay (DGD).

Over a small frequency range, a variation in frequency of $\Delta\omega$ will, primarily, simply cause a relative phase slippage of the two eigenmodes. In a time τ the slippage will be $\Delta\omega \cdot \tau$, so that, whenever this quantity is an integral multiple of 2π, we shall return to the same output polarization state. Hence, with varying frequency (over a small range) the output polarization state will vary cyclically, with a cyclical frequency difference given by

$$\Delta\omega_c = 2\pi/\tau \qquad (5.2b)$$

This is an important relationship because it provides an alternative method for the characterization of PMD.

However, real fibers do not retain uniform polarization properties over more than a few meters. There will be inhomogeneities owing to the manufacturing imperfections, variations in dopant concentrations, but, more especially in the case of installed telecommunications fiber, owing to bends, twists, and transverse pressures on the fiber. Consequently, the PMD analysis for the "long" fiber is rather different.

5.4.1.2 "Long" Fibers

For lengths of installed fiber in excess of about 10m the polarization properties of the fiber will not be uniform. The fiber, in this case, can be modeled as a concatenation of short lengths of fiber, each with its own, quasi-independent, polarization properties. When modeled in this way we shall see that the PMD will not now increase linearly with fiber length, as will now be explained.

Each section of fiber will be characterized by its polarization eigenmodes and the delay inserted between them by the length of the section and their velocity difference. Hence each section can be represented by a vector corresponding to the eigenmode diameter on the Poincaré sphere, of length equal to the delay (equivalent to a rotation of the Poincaré sphere about the eigenmode diameter). If the sections are infinitesimally small, and of equal length, the effect produced by any finite length of fiber is equivalent to an addition of these elementary vectors, because small rotations add vectorially. The resulting final value of PMD will be the amplitude of the resulting vector.

Let us represent each elementary vector by its components in three-dimensional space, (X, Y, Z). We assume that each vector has the same infinitesimal length, dl say, and that there are N of them, so that

$$N \cdot dl = L$$

where L is the length of the fiber. The nth section therefore has components (X_n, Y_n, Z_n), $n = 1$ to N.

Each of the $3N$ components is assumed to be randomly distributed about zero (the distribution being Gaussian in form), because we are assuming that each is independent of all the others, and that the polarization perturbations are randomly distributed along the fiber length, L.

The square of the amplitude of the resulting vector, formed by the addition of the elementary vectors will be given by

$$\tau^2 = \left(\sum_{n=1}^{N} (X_n) \right)^2 + \left(\sum_{n=1}^{N} (Y_n) \right)^2 + \left(\sum_{n=1}^{N} (Z_n) \right)^2$$

Since the components are independent, and is each equally likely to be positive or negative, it follows that over a long length of fiber the sum of the cross product terms will average to zero; that is,

$$\sum_{\substack{m,n=1\\ m\neq n}}^{N} X_m X_n = 0$$

and similarly for $Y_m Y_n$, $Z_m Z_n$.

Hence, for a long length of fiber,

$$\tau^2 = \sum_{n=1}^{N} \left(X_n^2 + Y_n^2 + Z_n^2\right) \tag{5.3}$$

Owing to the random distribution we shall also have

$$\langle X_n^2 \rangle = \langle Y_n^2 \rangle = \langle Z_n^2 \rangle = \alpha^2$$

where $\langle\ \rangle$ represents the average value.

Hence, from (5.3),

$$\tau^2 = 3N\alpha^2$$

Hence the value of τ will be given by

$$\tau = \sqrt{3N}\alpha \tag{5.4}$$

τ is the mean value of the amplitude of the final PMD vector, which we have identified as the PMD delay. From (5.4), and with $N.dl = L$, we see that

$$\tau = \left(\sqrt{3/dl} \cdot \alpha\right)\sqrt{L} \tag{5.5}$$

and the delay increases with the square root of fiber length, L.

5.4.2 Distinction Between "Long" and "Short" Regimes—Correlation Length

We have noted that fibers in the "long" and "short" regimes behave rather differently with regard to PMD, in particular, in the rate at which the PMD increases with length. It is clear that, for real fibers, we need some way of knowing in which regime the fiber lies. This is quantified via the *correlation length*, designated L_c, which, roughly speaking, is the length after which the polarization state of the light in the fiber has lost all knowledge of the input state, and is conditioned primarily by the polarization perturbations it has already encountered. Clearly, L_c is a statistical entity and must be defined, rigorously, in statistical terms, but it provides a convenient criterion for discriminating between the two regimes. For a length, l, of fiber,

$$l < L_c \quad \text{short regime}$$

$$l > L_c \quad \text{long regime}$$

The most convenient way of visualizing the concept of correlation length (as with so many other polarization problems) is to project the processes on to the Poincaré sphere. If we conceptualize a large number of equal lengths, l_e, of a given fiber, all subject to, statistically, the same perturbations, and then, with the same input polarization to each, we plot all the output polarizations on the Poincaré sphere (Figure 5.7) we see that when l_e is small, the output polarization states are all localized on the sphere (if the fiber were completely uniform over that length the output state, clearly, would be totally determined, and constant for all fibers: the output would be a single point on the sphere). As l_e grows for all the fibers, so the area occupied on the sphere grows in size until, at some value of, the sphere is uniformly covered by all possible output states. At this point the output state has become essentially independent of the input state, since all output states are equally likely. The value of l_e for which this first happens is the correlation length, L_c.

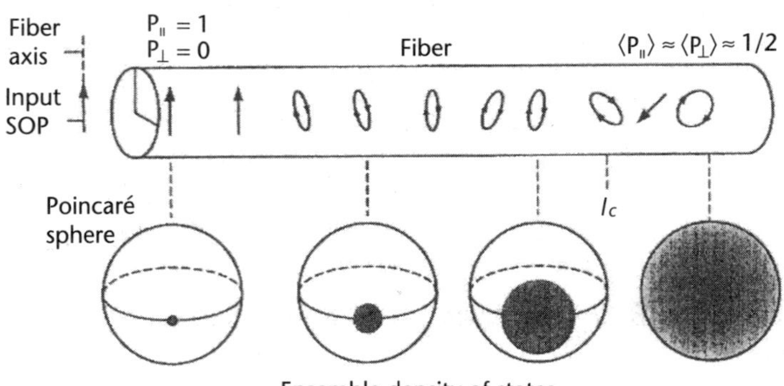

Figure 5.7 Decorrelation of polarization states in a long fiber. (*From:* [1]. © 1997 Academic Press. Reprinted with permission.)

Formally, L_c is defined as the length of fiber for which leakage of power from the input polarization state into its orthogonal state is 0.135 (i.e., $1/e^2$) of the total input power.

However, it is not easy to measure the value of L_c as it has been defined in the preceding paragraph, for it would require that the fiber be cut up into a large number of sections of varying length! In order to measure it satisfactorily, however, we can use the equivalence of a variation in τ with a variation in the optical frequency of the input light given in (5.2b). For a fiber in the short regime, a variation in frequency will cause the output state to rotate deterministically around the eigenvector on the Poincaré sphere; for one in the long regime, the effect is to vary, statistically, all the τs of the concatenated elements, thus effectively doing the same job as moving to a fiber subject to perturbations obeying the same statistics. It follows, therefore [1], that if a frequency variation causes the output states to be distributed uniformly over the surface of the Poincaré sphere, the fiber is in the long regime, and vice versa.

5.4.3 Formal Analysis of PMD

The fairly simplistic approach to PMD which we have adopted up to this point is useful in obtaining a physical "feel" for the phenomenon, but, in order to determine quantitatively what is the effect of PMD on operational fiber communications systems, we require a more rigorous approach to the subject, using the polarization transfer function as embodied in the Jones vectors.

We have already seen in Section 3.11.4 that any polarization element (uniform or nonuniform) can, with suitably chosen axes, be represented by a Jones matrix of the form:

$$\mathbf{M} = \begin{pmatrix} \alpha + i\beta & -\gamma + i\delta \\ \gamma + i\delta & \alpha - i\beta \end{pmatrix}$$

and that this can be written in the form:

$$\mathbf{M} = \begin{pmatrix} u_1 & u_2 \\ -u_2^* & u_1^* \end{pmatrix}$$

where the asterisk denotes the complex conjugate. Hence any input polarization vector, $\mathbf{P}$, will be acted upon by the matrix to yield an output polarization vector $\mathbf{Q}$:

$$\mathbf{Q} = \mathbf{M} \cdot \mathbf{P}$$

or, in more explicit form:

$$\begin{pmatrix} Q_x \\ Q_y \end{pmatrix} = \begin{pmatrix} u_1 & u_2 \\ -u_2^* & u_1^* \end{pmatrix} \begin{pmatrix} P_x \\ P_y \end{pmatrix}$$

$P_{x,y}$, $Q_{x,y}$ are, of course, also complex quantities, since they embody both the amplitudes and phases of the respective components. Now **M** is a unitary matrix with orthogonal eigenmodes (in the absence of polarization-dependent loss or gain (PDL/G)) and these eigenmodes are those states of polarization which are the same at the output as at the input. These modes will travel at different velocities and would serve to define the PMD of a nonuniform fiber length by virtue of the time delay imposed between them as a result of the propagation through the fiber. However, this is only the case for a single optical frequency and in the absence of wavelength dispersion; for, from (5.2a), we note that the time delay between the eigenmodes is frequency-dependent, this being the consequence of a frequency dependence of the differential refractive index, Δn. This means that, when a spread of frequencies is present (as there always will be from any real optical source) there will be a mixture of eigenmodes and phase delays. When viewed on the Poincaré sphere, the result is that, for any given input polarization, as the frequency is varied, the output polarization rotates about a certain eigenvector, $\mathbf{q}_+$ $\mathbf{q}_-$ in Figure 5.8. Just as in the case of the short, uniform fiber, the rotation is complete after a characteristic frequency shift $\Delta\omega$. If the process is repeated for other input polarizations the rotation is repeated about the same vector with circles of different radii (see Figure 5.8).

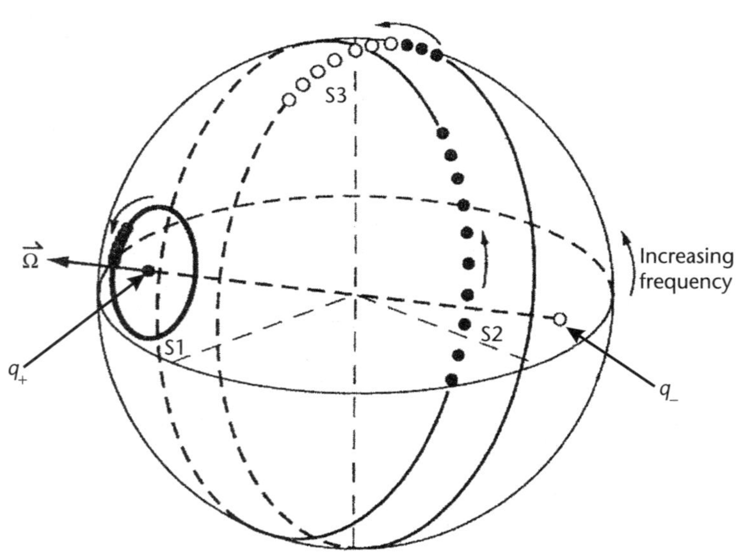

Figure 5.8 Rotation of output states about a particular eigenvector for varying optical frequency. (*From:* [2]. © 1988 IEEE. Reprinted with permission.)

The obvious question now to ask is what is the physical significance of the vector $\mathbf{q}_+ \, \mathbf{q}_-$?

Looking again at Figure 5.8, the supplementary question that might be asked is whether it is possible to choose an input polarization state such that the output polarization state is q_+ or q_-. Common sense tells us that it must indeed be possible, because otherwise there would be a discontinuity in the choice of output states, and thus also of input states, as the radius of the circle decreases, and this could not have any physical meaning. Consequently, we confidently expect that there will be two input polarization states that have, respectively, q_+ and q_- as their output states. And the really important point is that these output states will be independent (to first order) of optical frequency. This, in turn, means that all the frequencies, in a fairly small range, from an optical source will arrive in the same polarization state, and thus will not have suffered any relative phase delay. (This state of affairs can be regarded as analogous to the *zero-dispersion* wavelength for a chromatically dispersive medium.) Hence, we can summarily formalize the position by stating that, at any optical frequency, there exist two input and two corresponding output polarization states with the property that the output states are invariant (to first order) with changes in optical frequency. The two output states are orthogonal (in the absence of PDL/G) and, for a pulse launched into either one of the states, all frequencies arrive with the same phase and polarization state, so that distortion, and thus PMD, is minimized. The two states are called the *principal states of polarization* (PSPs) of the fiber and they will possess different group velocities. The effective PMD delay is now the difference in arrival time for the two states, assuming that the input light has launched components in each of the two corresponding input states. It will not be possible to mitigate PMD by fixing the input state to correspond to one of the required states for minimum PMD in any practical installed system, because temperature and other environmental variations will cause variations in the principal states with time.

The principal states vector $\mathbf{q}_+ \, \mathbf{q}_-$ now conveniently allows us to define a PMD vector, $\mathbf{\Omega}$. This is a vector on the Poincaré sphere with direction $q_+ - q_-$ and magnitude equal to the DGD (τ) between the PSPs.

These ideas were first recognized and formalized mathematically by Poole and Wagner [3] by their writing of the Jones matrix with frequency-dependent elements,

$$\mathbf{M} = \begin{pmatrix} u_1(\omega) & u_2(\omega) \\ -u_2^*(\omega) & u_1^*(\omega) \end{pmatrix}$$

and deriving the condition necessary for the input states to yield zero frequency dispersion for the output states. The matrix manipulation gives the result that the propagation delay between the principal states is given by

$$\tau = 2\left(\left|du_1/d\omega\right|^2 + \left|du_2/d\omega\right|^2\right)^{1/2} \tag{5.6}$$

where | | denotes the modulus of the complex number. τ now represents the PMD for the nonuniform fiber.

These ideas have been convincingly demonstrated by observation of the output from a pulse launched into each of the two input principal states (Figure 5.9).

Hence the principal states model of a nonuniform fiber can be regarded as the analog of the eigenmodes model for a uniform fiber. It is a powerful and convenient tool for description and manipulation of PMD phenomena in installed-system communications fibers.

5.4.4 The Statistics of PMD in Installed Fibers

As has already been noted, the PMD properties of long lengths of installed fiber are not constant with time. The variations are due, for example, to changes in temperature, the effect of wind on exposed cables, and earth movements in buried cables. The result is that the measurement of PMD over a long length of fiber must be treated statistically. It is fairly straightforward to derive the first-order statistics of the variation of the PMD from the ideas described in Section 5.4.1.

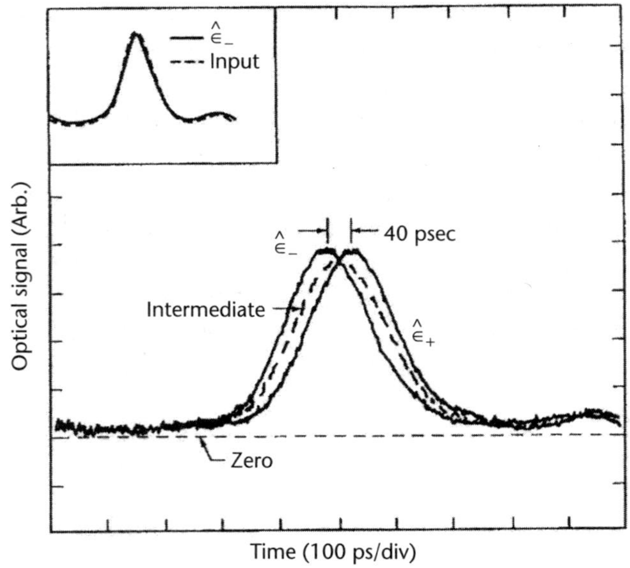

Figure 5.9 Direct observation of the delay between principal states. (*From:* [4]. © 1988 Optical Society of America. Reprinted with permission.)

Considering, again, the fiber to be composed of a large number of concatenated birefringent elements, each described by a polarization eigenvector, we have seen that the Cartesian components, X, Y, Z, of each eigenvector, are randomly distributed according to Gaussian statistics; that is,

$$P(X, Y, Z) = A \exp(-X^2, Y^2, Z^2/2\sigma^2)$$

where A and σ are constants, and $P(X)\,dX$ is the probability of finding X in the range X to $X + dX$; and similarly for Y and Z.

We wish to determine the probability of the PMD having a given, final value of τ at the end of the fiber.

The probability of any given set of components, $X_1, X_2, \ldots, X_n$, of the concatenating birefringence vectors is

$$p(X_1, X_2, \ldots X_N) = A^N \exp\left(-X_1^2/2\sigma^2\right) \exp\left(-X_2^2/2\sigma^2\right) \ldots \exp\left(-X_N^2/2\sigma^2\right)$$

and similarly for $p(Y_1, Y_2, \ldots, Y_N)$ and $p(Z_1, Z_2, \ldots, Z_N)$.

Hence the combined probability of a full set of given X, Y, Z components is

$$\begin{aligned} p(X_1, X_2, \ldots X_N;\, Y_1, Y_2, \ldots Y_N;\, Z_1, Z_2, \ldots Z_N) \\ = A^{3N} \exp\left(-\sum_{n=1}^{N} \left(X_n^2 + Y_n^2 + Z_n^2\right)/2\sigma^2\right) \\ = A^{3N} \exp(-\tau^2/2\sigma^2) \end{aligned} \qquad (5.7)$$

from (5.3).

Now the probability for a given final value of τ must include only those combinations of the X_n, Y_n, Z_n which can give rise to that value of τ, and this can be derived by multiplying their probability of occurrence by the probability that their final value of τ will be within the volume which lies between τ and $\tau + d\tau$ on a Cartesian plot of all values of X, Y, Z (see Figure 5.10). This volume is, clearly, $4\pi\tau^2 d\tau$, so that the final probability for a given value of τ, from (5.7), is

$$p(\tau) = 4\pi\tau^2 A^{3N} \exp(-\tau^2/2\sigma^2)$$

or

$$p(\tau) = B\tau^2 \exp(-\tau^2/2\sigma^2) \qquad (5.8)$$

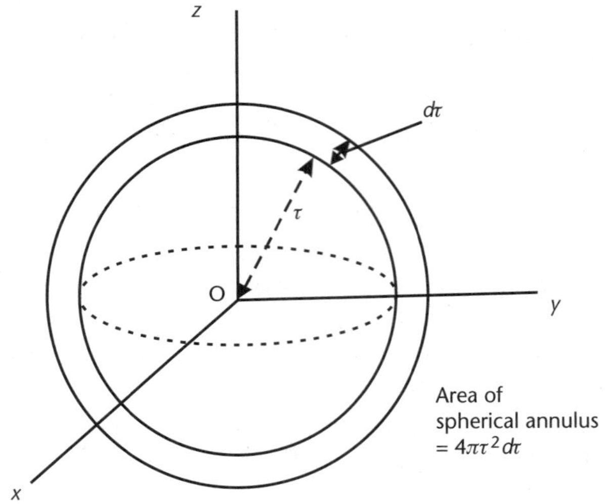

Figure 5.10 Volume of annulus for density of probability calculation.

This form of $p(\tau)$ is known as a Maxwell distribution and an example is shown in Figure 5.11.

From this distribution, once B and σ are known for a given fiber, it is clear that one can calculate the probability that τ will be greater than any threshold value, and it is this information which is required in order to determine what will be the effect on the performance of a communications system which uses that length of fiber.

Since there is a probability of unity that one value of τ will occur, it follows that

$$\int_0^\infty p(\tau)\, d\tau = 1$$

which imposes certain constraints on B and σ.

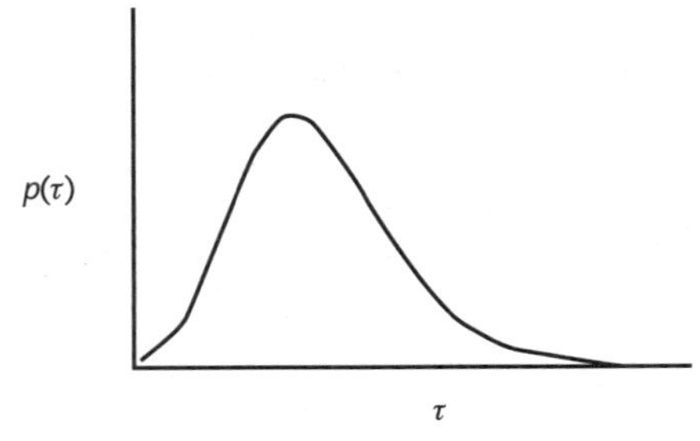

Figure 5.11 The Maxwell probability distribution for $p(\tau)$.

Also, $\langle\tau\rangle$ is the mean value of over the Maxwell distribution, where

$$\langle\tau\rangle = \int_0^\infty \tau p(\tau)\, d\tau$$

In other words, it is the averaged value of a given value of τ multiplied by the probability (frequency) of its occurrence.

B and σ can now be related by [3]

$$B = \sqrt{2/\pi}/\sigma^3$$
$$\sigma = \sqrt{\pi/8} \cdot \langle\tau\rangle$$

Hence, a measure of $\langle\tau\rangle$, the average value of τ, will determine the parameters, B, σ, of the distribution.

We also can relate $\langle\tau^2\rangle$ with $\langle\tau\rangle$ for the Maxwell distribution:

$$\langle\tau\rangle^2 = 8/3\pi \cdot \langle\tau^2\rangle$$

Hence the requirement for any measurement of the effect of PMD in long fibers is to measure this mean value $\langle\tau\rangle$ and, in the next section, we shall describe some ways of doing this.

In order to define quantitatively the effect that a given level of PMD will have on communications system performance in a digital system, we must consider the effect of the broadening of the pulses on the error rate. As we have seen, an error occurs when a "0" is interpreted (by the detection system) as a "1," or vice versa (Figure 5.12). The errors are caused by random noise, unavoidably present in any electronic system, which causes the signal at certain times to cross the threshold set to determine whether any given slot contains a "0" or a "1" (see Figure 5.12). Provided that the statistics of the noise are known, it is possible to predict how often this will happen, and thus to determine the bit-error rate (BER) for the communications system. Clearly, when dispersion is present the effect of the noise is enhanced because the amplitude of the bits is reduced, thus reducing the signal-to-noise ratio (SNR) (see Figure 5.12) and increasing the BER. The extent to which this is allowed to happen must be arbitrary, depending on how reliable the communications link is required to be, but, for long-haul telecommunications systems, the commonly accepted criterion is that the BER should not exceed 10^{-12} for more than 30 minutes per year. (A period of time for which the BER exceeds 10^{-12} is termed an

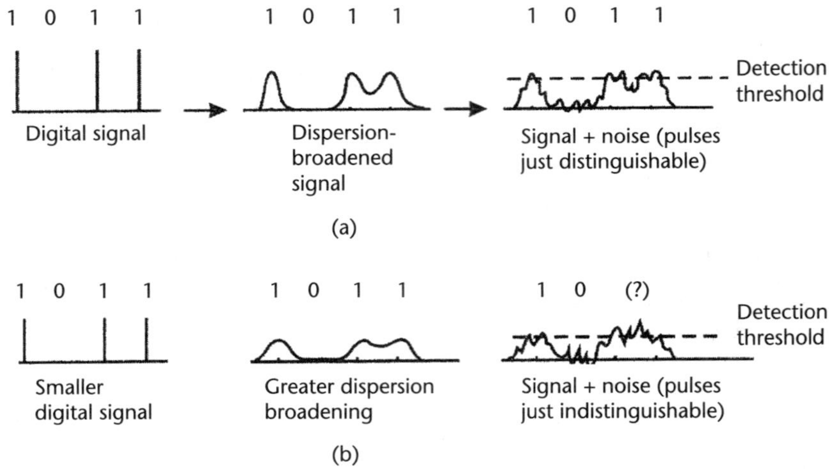

Figure 5.12 Digital errors caused by dispersion in the presence of noise: (a) effect of dispersion + noise: pulses just distinguishable; and (b) effect of dispersion + noise: pulses indistinguishable.

outage.) When this criterion is considered in relation to PMD statistics, it translates into a requirement that $\langle\tau\rangle$ does not exceed 14% of the bit period (T); that is,

$$\langle\tau\rangle < 0.14/W = 0.14T$$

where W is the system bit-rate.

5.4.5 Measurement of PMD

There are many methods for measurement of PMD depending on the required accuracy, speed of measurement, and cost. Just a few of these will be described here, to give the flavor of the possible approaches.

For the short-length regime the problem is quite straightforward, for the fiber is uniform and its properties are static. From (5.2b):

$$\tau = 2\pi/\Delta\omega_c$$

We note that τ can be measured via the optical frequency change necessary to cause the output polarization state to perform a complete cycle, back to its initial state, as the frequency change alters the phase between of the eigenmodes. Hence, what is required here is a frequency-agile source, a polarization element to control the input state and a polarization analyzer to determine the output state. For the long-length regime it must be remembered that the PMD is varying statistically, so that we need to measure the average value, $\langle\tau\rangle$. Because the statistics are known from (5.8), this measurement then accurately quantifies the PMD. In order to vary statistically without resorting to extensive temperature conditioning of large-scale changes in the geometrical configuration of the fiber, the stratagem adopted is that of sweeping the optical frequency. Both experimental and theoretical studies [5–7] support the assumption that the variation in phase between the principal states occasioned by this is statistically equivalent to the variations which are caused by the environmental factors.

Two methods for measuring $\langle\tau\rangle$ in long fibers will now be described.

5.4.5.1 The Jones Matrix Method

In this method [8] the full Jones matrix is measured as a function of frequency (Figure 5.13). For a given input polarization the output polarization state from the fiber is measured for a number of closely spaced frequencies. The input polarization state is then changed and the process repeated. The frequency-dependent Jones matrix can then be computed. At least three input states are normally employed. This method is a complete method in that it specifies the frequency-dependent Jones matrix at any given center frequency, thus allowing $\langle\tau\rangle$ to be measured to both first and higher orders (i.e., it also measures the frequency dependence of $\langle\tau\rangle$). By sweeping the frequency over a wider range, the statistical variation of $\langle\tau\rangle$ can be induced and its value measured. The main disadvantage of this method lies in its sophistication. A precise, tunable source is required as is an accurate polarization analyzer; the computation also is time consuming.

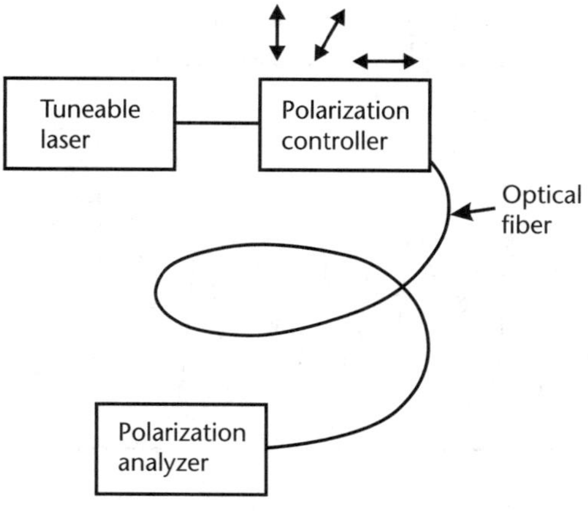

Figure 5.13 Setup for the Jones matrix method for PMD measurement.

5.4.5.2 The Fixed Analyzer Method

In this method, again, a frequency tunable source is required, but this time just a single input polarization state is used with a fixed polarization-state analyzer [9] (Figure 5.14). As the frequency is varied the output signal level varies quasi-sinusoidally as shown in the figure. This is, again, the result of the varying phase between the principal states passing through large multiples of 2π.

It can be shown [9] that the mean value, $\langle\tau\rangle$, is given by

$$\langle\tau\rangle \approx k_1 N_m / \Delta\omega$$

where N_m is the number of times the signal level crosses its mean level in the frequency interval, $\Delta\omega$. The larger is N_m, the better the accuracy of measurement for $\langle\tau\rangle$. Alternatively, one can count the number (N_e) of maximum and minima in a given frequency interval. In this case,

$$\langle\tau\rangle \approx k_2 N_e / \Delta\omega$$

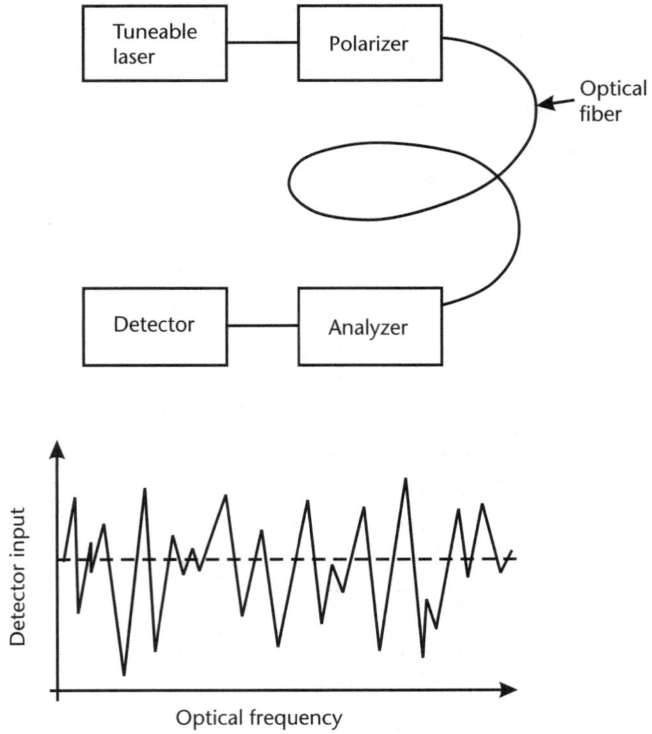

Figure 5.14 The fixed analyzer method for PMD measurement.

An interesting feature of this method is that it provides an indication of whether the fiber under test (FUT) lies in the short-length or the long-length regime for

$$N_e/N_m = 1.55 \qquad \text{for } L \gg l_c$$

$$N_e/N_m = 1 \qquad \text{for } L \ll l_c$$

The advantages of this method are that it is quick to set up and analyze. The disadvantages are that it can only measure $\langle \tau \rangle$, not τ, and that N_e and N_m must be large (>50) for accurate measurement.

5.4.5.3 Measurement of the PMD Distribution

Several methods have been used to measure the spatial distribution of PMD along a long length of fiber. This is very often extremely useful for the identification of particular sections of fiber in the length which have an anomalously high value of PMD resulting from local environmental changes (e.g., Earth movements, faulty fastenings, mechanical diggers, and so forth) and which are resulting in unacceptably high overall link PMD. These methods usually involve the launch of a high-power pulse into the fiber and then polarization analysis of the light which is continuously Rayleigh-backscattered, back to the launch end, as the pulse propagates (Figure 5.15). The generic method is known as polarization-optical time domain reflectometry (POTDR) [10]. In this, the light returning to the launch end at time t has performed a go-and-return passage up to distance in the fiber where

$$z = c_g t/2$$

c_g being the group velocity of the pulse in the fiber. Hence, its polarization state at time t is a measure of the polarization properties of that double passage through a length, z, of fiber.

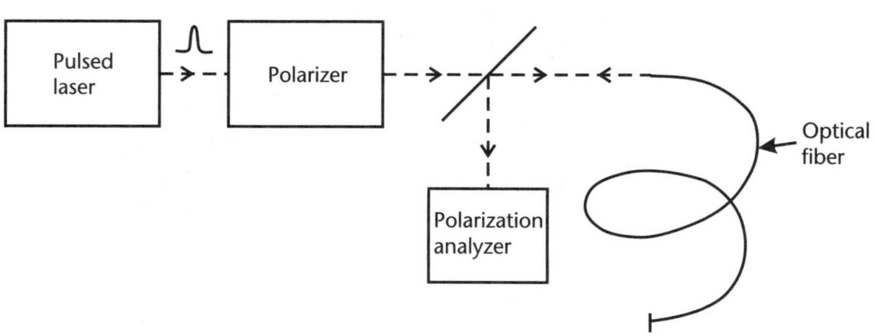

Figure 5.15 POTDR setup.

We have already noted (Section 3.10) that any lossless, reciprocal polarization element can be modeled as an equivalent retarder/rotator pair (Figure 3.16). Suppose that the length, z, of fiber is so modeled, and that the retardation of the equivalent retarder is $\delta_e(z)$, its orientation (with respect to some arbitrary reference direction) is $q_e(z)$, and that the rotation of the equivalent rotator is $\rho_e(z)$. For a reciprocal element the rotation of the rotator will be cancelled on return passage so that, in backscatter, the element will appear as a pure retarder of retardation $2\delta_e(z)$ and orientation $q_e(z)$.

Consider, now, the infinitesimal element of fiber beyond the end of the element of length, z, and of width dz.

Suppose that this element has polarization properties characterized by the actual (not equivalent) birefringences, $\delta(z)$, $q(z)$, $\rho(z)$. The values for these quantities determine how quickly $\delta_e(z)$, $q_e(z)$, $\rho_e(z)$ change, so that relations can be established between the two sets [11]:

$$\delta(z) = ((\partial\delta_e/\partial z)^2 + \sin^2 2\delta_e(\partial q_e/\partial z)^2)^{1/2} \tag{5.9a}$$

$$q(z) = q_e + \rho_e + 1/2 \cdot \tan^{-1}(\sin 2\delta_e(\partial q_e/\partial z)/(\partial\delta_e/\partial z)) \tag{5.9b}$$

$$\rho(z) = \partial\rho_e/\partial z + 2\sin^2 \delta_e(\partial q_e/\partial z) \tag{5.9c}$$

Now the element up to a point, z, in the fiber can be represented by the Jones Matrix:

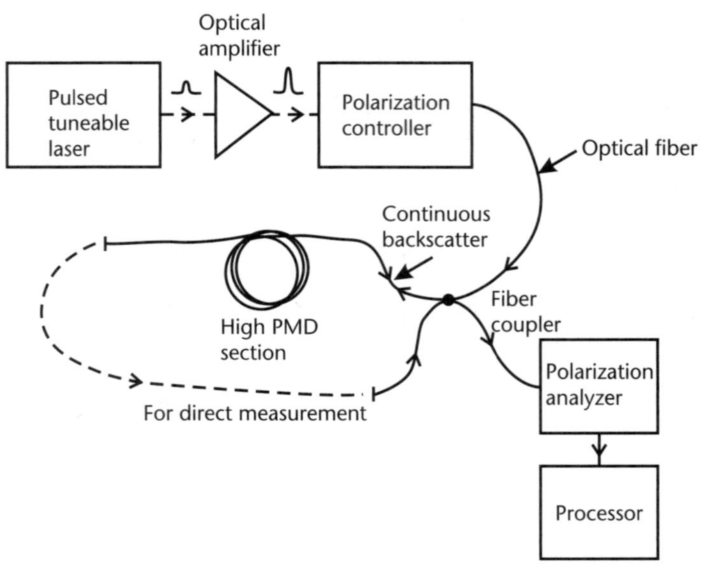

Figure 5.16 Experimental setup for PMD distortion measurement.

$$\mathbf{M}(\mathbf{z}) = \begin{pmatrix} M_{11} & M_{12} \\ -M_{12}^* & M_{11}^* \end{pmatrix}$$

where

$$M_{11} = \cos\,\delta_e \cos\,\rho_e + i\,\sin\,\delta_e \cos(2q_e + \rho_e)$$
$$M_{12} = \cos\,\delta_e \sin\,\rho_e - i\,\sin\,\delta_e \sin(2q_e + \rho_e)$$

and the PMD, represented by the delay between the principal states, is given by [3]

$$\tau = 2(|\partial M_{11}/\partial\omega|^2 + |\partial M_{12}/\partial\omega|^2)^{1/2}$$

giving

$$\tau(z) = 2 \cdot \lambda/\omega \cdot ((\partial\delta_e/\partial\lambda)^2 + (\partial\rho_e/\partial\lambda)^2 + 4\,\sin^2\,\delta_e \cdot (\partial\rho_e/\partial\lambda + \partial q_e/\partial\lambda)(\partial q_e/\partial\lambda))^{1/2}$$

$\tau(z)$ can, therefore, be measured provided that we can measure δ_e, q_e, ρ_e as a function of both z and λ. Figure 5.16 shows an experimental arrangement for doing this employing, again, a wavelength-tunable, pulsed source of light.

The only difficulty now, of course, is that ρ_e is not measurable. However, $\partial\rho_e/\partial\lambda$ is measurable using (5.9b) and making use of the fact that $q(z)$, being a physical direction in space, is independent of λ. Hence, differentiating (5.8b) with respect to λ and setting $\partial q(z)/\partial\lambda = 0$, we find

$$\partial\rho_e/\partial\lambda = -\partial q_e/\partial\lambda - 1/2(\partial/\partial\lambda(\tan^{-1}(\sin\,2\delta_e(\partial q_e/\partial z)(\partial\delta_e/\partial z))) \tag{5.10}$$

Hence all quantities on the RHS of (5.10) are now measurable and $\tau(z)$ can be computed.

Figure 5.17 shows the result of one such computation for a 5-km length of fiber. The total, "straight-through" PMD value could also be measured directly, for comparison (see Figure 5.16).

This technique thus allows an effective identification of sources of anomalous PMD in telecommunications systems. Other techniques for performing this task can be found in the literature, for example, [12, 13].

5.4.6 Compensation for PMD

Clearly, the best way to compensate for the effects of PMD is to manufacture polarization-transparent optical fiber, and manufacturing procedures certainly

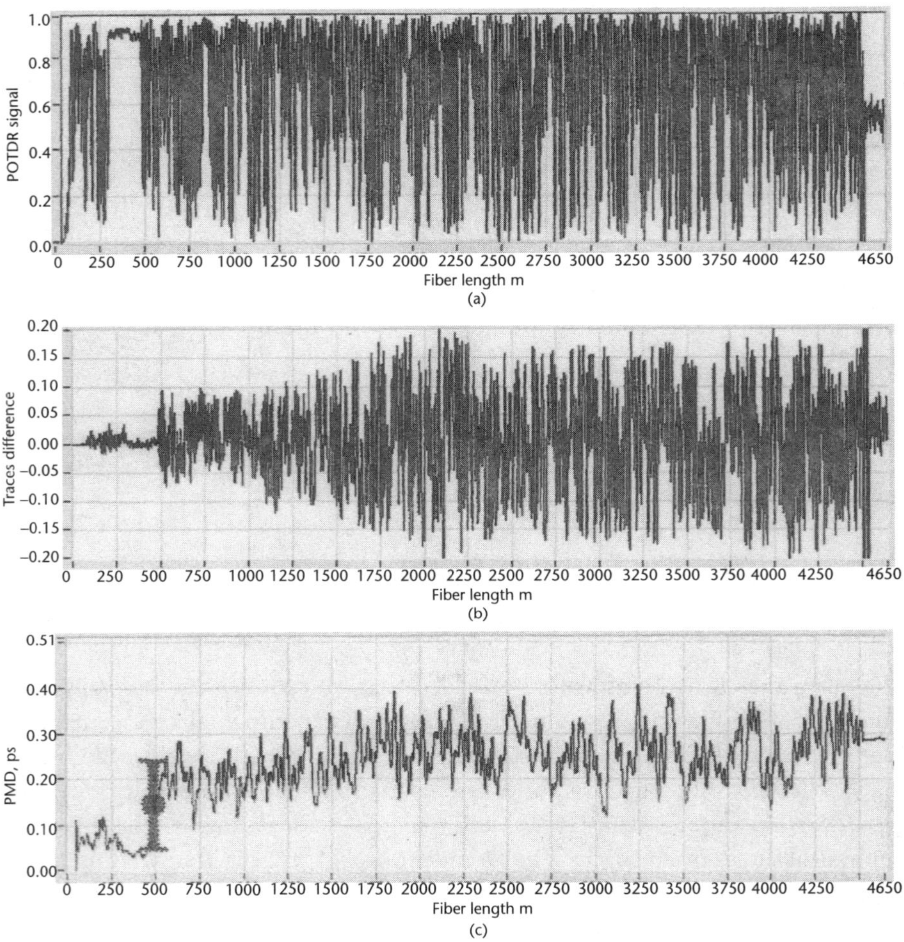

Figure 5.17 Experimental results for a distributed PMD measurement: (a) typical POTDR trace; (b) difference in traces for two wavelengths; and (c) comparison of distributed and direct measurements of PMD. (*From:* [13]. © 2006 IEEE. Reprinted with permission.)

have provided greatly improved fibers in this regard, with PMD specifications of <0.1 ps.km$^{-1/2}$. However, with this figure and, inevitably, externally imposed PMD from bends, twists, and fiber components in the system, the implication for long haul systems several hundred kilometers in length is that PMD still will be a problem for bandwidths of 40 Gb $\cdot$ s^{-1} and above. Consequently, a number of schemes for compensating for the effects of PMD have been studied: this whole area is often termed *PMD mitigation*. There are essentially two types of approach to PMD mitigation: optical compensation and electronic compensation. We shall provide here just one example of each approach, to indicate the broad methodology. Other examples may be found in the literature, for example, [14, 15].

5.4.6.1 Optical Compensation

Optical compensation methods for PMD essentially comprise a manipulation of the emerging polarization state of the signal power in order to reduce the effect which PMD has on the light, and hence on the system error rate. To do this requires polarization components whose properties are controlled by a feedback error signal, and the feedback control, clearly, must be effected in a time which is small compared with the changes in the PMD characteristics of the system. Hence a compromise must be made between speed and sophistication, the latter being identifiable with the level of compensation likely to be achievable. The first level of sophistication refers to the "order" of the compensation: "first order" refers to the compensation at the center wavelength only; "second order" to the first derivative of the PMD with respect to frequency, and so on. There are many approaches to the solution of this complex problem, and all that we shall do here is give one example which illustrates the general philosophy employed.

The example, which is of a first-order arrangement, is shown in Figure 5.18. Here, the compensating components are a rotator followed by a retarder. The rotation and the retardation are controlled by error signals from the optical detector/processor. The principle by which this compensation operates is shown in Figure 5.19(a). The PMD vector for the fiber in the communication system is represented by $\mathbf{\Omega}_F$, while that for the rotator/retarder pair is $\mathbf{\Omega}_C$. If the polarization state of the light launched into the fiber at the front end is represented by the vector $\mathbf{P}$, then the requirement is for $\mathbf{P}$ to be a principal state of the combined system, fiber plus compensator. This can be achieved if the total PMD vector, $\mathbf{\Omega}_T$, the vector sum of $\mathbf{\Omega}_F$ and $\mathbf{\Omega}_C$, aligns with $\mathbf{P}$ [Figure 5.19(b)]. In order to achieve this alignment, $\mathbf{\Omega}_C$ must be swung around the Poincaré sphere, so that we need two adjustments, one for azimuth (φ), and the other for elevation (θ), in order to minimize the total angle, ψ, between $\mathbf{P}$ and $\mathbf{\Omega}_T$.

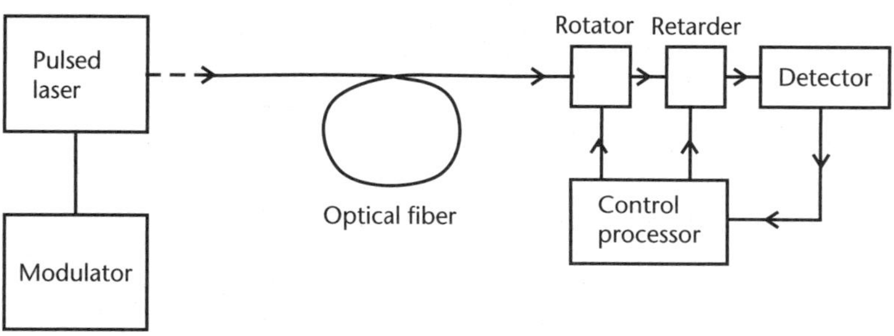

Figure 5.18 Basic arrangement for first-order PMD compensation.

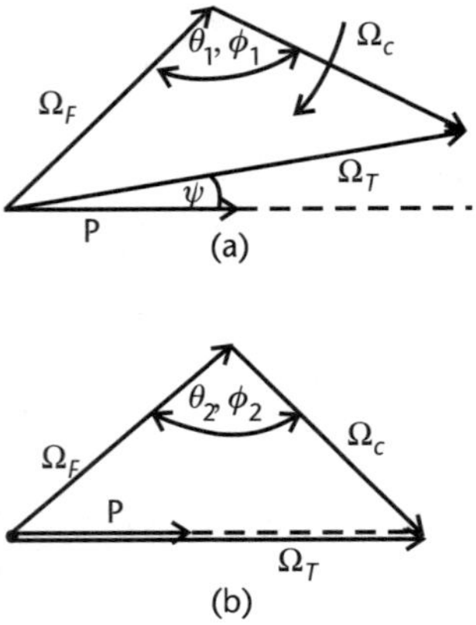

Figure 5.19 (a, b) Vector manipulations for optical compensation.

First, we require an error signal to control the azimuth by a feed to the rotator. This can be obtained by measuring the mean degree of polarization (DOP) of the output signal. Because the uncompensated signal essentially consists of the two principal states displaced in time, the polarization state will vary within the pulse length, hence effectively depolarizing the averaged state. The DOF can be measured by splitting off a portion of the signal and measuring its Stokes parameters. A voltage level proportional to the DOP can then be applied to the rotator, which will rotate the polarization state of the light to maximize the DOP: this will swing $\mathbf{\Omega_T}$ from a point A, say, to B, on the Poincaré sphere (Figure 5.20), which is as close to P as a pure rotation can

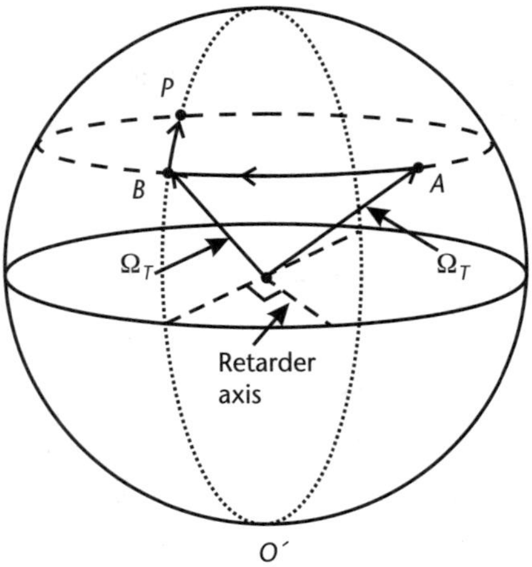

Figure 5.20 Poincaré sphere representation of the action in first-order PMD compensation.

achieve. B and P will lie on a circle of longitude. The next rotation should be about an equatorial axis normal to the plane of that circle so that B can rotate up to **P**. The error signal for this can be obtained from the modulation spectrum of the uncompensated signal. Because the PMD splits the signal power into the two PSP components, the modulation component at frequency $1/2\tau$ (where τ is the delay between the PSPs) will be reduced in amplitude, because it will average to zero over the complete pulse length. Hence the feedback loop is set so as to provide that this amplitude is maximized by the setting of the retarder. However, the required adjustment of the retarder should be in both its orientation and the retardation, and there is only one error signal for these adjustments in this two-stage system, so that the retarder does the best it can, and this is followed by another rotation, and so on. This means that the required adjustment is not achieved immediately, and the system will hunt to find the best achievable position. Additionally, there is the problem, common to all feedback control circuits, of the system getting trapped into a *local* rather than a *global* optimal point, because of the complexity of the relationship between the compensated signal and the feedback signals. This is one aspect of the more general problem of arranging suitable software algorithms for use of the feedback signals for effective control of the polarization manipulations required for useful mitigation. This is a taxing software task.

A typical improvement in the BER brought about by such an arrangement is shown in Figure 5.21, where the improvement is expressed in terms of the variation in the communications system's error rate with respect to the bit period (clearly, the greater the bit period, the greater the allowable DGD for a given error rate).

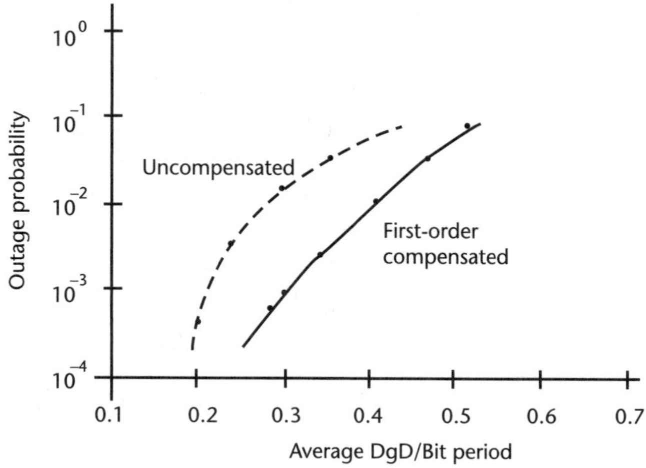

Figure 5.21 Improvement in outage rate affected by first-order PMD compensation.

Such an arrangement is, of course, limited by its first-order nature, and will not allow for variations in the PSPs with frequency.

A variety of other first-order and second-order schemes has been devised. Always there will need to be a compromise between effectiveness and complexity. As complexity increases, so does cost, and also the difficulty of ensuring that the feedback loop acts sufficiently quickly to track the statistical changes in the PMD vector.

5.4.6.2 Electronic Compensation

Electronic compensation for PMD concerns itself with the processing of the post-detection signal. This usually involves sensing the distortion introduced by PMD into the waveform of the digital pulse stream, and using distortion-characterizing parameters to feed back to shaping circuits prior to the final output.

An example of this is shown in Figure 5.22, where the distortion analyzer is acting back on to a three-way power splitter with variable splitting ratios [16]. The different channels then suffer variable delays, and are then recombined to constitute the final output signal. Clearly, the criterion employed in this feedback processing will be that of minimizing the distortion.

Research in the area of PMD mitigation, at both the optical and electronic levels, is continuing actively.

5.5 Coherent Optical Communications Systems

The optical communications systems which have been discussed so far have all been of the type called, in conventional parlance, amplitude modulation (AM) systems. By this it is meant that the power level of the source is varied in sympathy with the signal modulation. Since an optical source is more readily categorized by (among other things) its output power, in the case of optical communications we should perhaps speak of *power modulation* or, since the

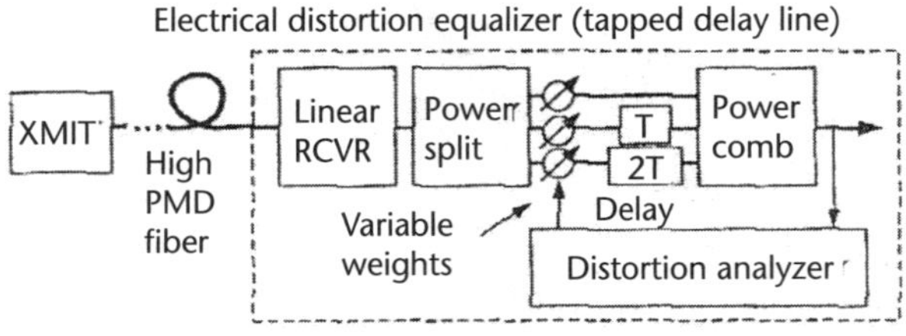

Figure 5.22 An electronic compensator for PMD. (*From:* [6]. © 1990 IEEE. Reprinted with permission.)

power propagates in a fiber of fixed cross-section, *intensity modulation* (IM), rather than amplitude modulation. The receiver in this case has a relatively simple task: to provide an electrical current proportional to the power it receives. It is a *direct detector* (DD) and such systems are often referred to as IM/DD systems.

Such systems have the important advantage of simplicity, and require relatively unsophisticated (and therefore cheap and reliable) components. However, they also have a number of disadvantages.

The first is that they are relatively insensitive. This means that it is difficult to obtain good receiver signal-to-noise ratios (SNRs) over long distances, thus necessitating frequent amplifying repeaters. The reason for the insensitivity is that the optical signal level is small after a long distance transmission, and the quantum noise is such as to allow a maximum SNR (see Section 1.7):

$$\text{SNR} = \left(\frac{P_s}{Bh\nu}\right)^{1/2} \tag{5.11}$$

P_s being the received signal power, B the bandwidth, h the quantum constant, and ν the optical frequency. In other words the smaller the SNR is, the smaller the received power is. Any subsequent amplification in the detection system can only degrade the SNR further, since it will add noise.

The second disadvantage is their large optical bandwidth. A typical multimode semiconductor laser has a spectral width ~ 5 nm, which corresponds to ~ 1,000 GHz of bandwidth. Since spacing between channels in a wavelength-division-multiplexed (WDM) system needs to be ~ 10 times the source width (to avoid cross-talk between channels), this means that each channel effectively occupies ~ 10,000 GHz. Such a large frequency spread does not allow more than one or two channels in either of the 1,300- and 1,550-nm transmission "windows" in silica fiber (Figure 2.14).

The ~ 1,000-GHz spectral width of the multimode semiconductor laser means that it is little more than an optical noise source (in communications terms) which, in digital AM systems, is simply switched on and off.

The development of radio and microwave techniques has shown that, with spectrally pure sources, much better system performance can be achieved by modulating the frequency, phase or polarization state, rather than simply the amplitude. In order to do this, the modulation parameter (e.g., frequency in FM) must be stable to better than ~ 1% of the modulation bandwidth if the signal is not to be distorted. Hence the requirement is for high power sources with narrow line-width, and good frequency/phase stability. Such sources are clearly going to possess a high degree of coherence, and the systems based on them are thus known generally as *coherent* systems.

Let us examine some of these ideas in the context of optical-fiber communications.

Suppose that we were to have available a "pure" high power optical source (i.e., a laser) whose output at the far, receiving, end of the fiber could be characterized in terms of its optical electric field as, effectively, a pure sinusoid:

$$E_s = e_s \cos(\omega_s t + \varphi_s) \tag{5.12}$$

Suppose, first, that this source is used in an IM/DD system with the information signal directly modulating the mean optical power level.

When this signal is directly detected, the detector provides an output current proportional to the input optical power (Section 1.9). The optical power is proportional to the intensity of the wave; that is, to the square of its amplitude, averaged over the response time of the photodetector:

$$P_s = C\langle e_s^2 \cos^2 (\omega_s t + \varphi_s)\rangle = \frac{1}{2} Ce_s^2 \langle (1 + \cos 2(\omega_s t + \varphi_s))\rangle$$

where C is a constant.

Now since the detector cannot respond to frequencies as high as $2\omega_s$, the current that flows is proportional only to the first term:

$$i_d \sim P_s = \frac{1}{2} Ce_s^2 \tag{5.13}$$

The current will therefore follow a modulation of the power level of the source, $P_s(t)$, up to the maximum speed of response of the photodetector.

Suppose now that we have available yet another pure optical source but at a different optical frequency. Let us describe its output by

$$E_L = e_L \cos(\omega_L t + \phi_L) \tag{5.14}$$

Let us allow the two sources described by (5.12) and (5.14) to fall simultaneously on to the photodetector (Figure 5.23). Now the input power and the resulting output current will be proportional to the time-averaged value of the square of the total electric field of the two waves, so that, *assuming that they have the same polarization*, the optical power will now be represented by

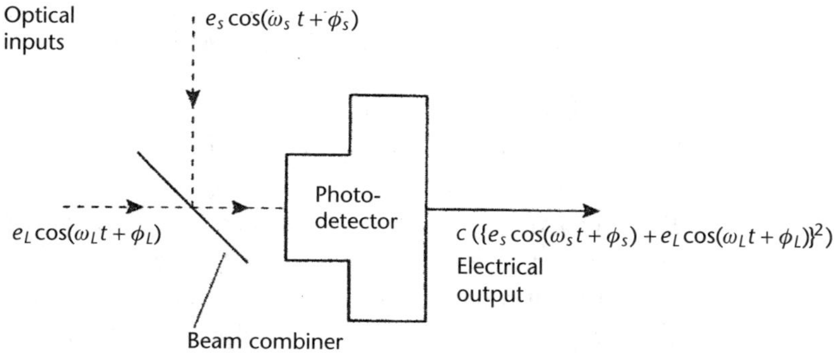

Figure 5.23 Square-law detection mixing of two optical signals.

$$
\begin{aligned}
P_D &= C\langle (E_s + E_L)^2 \rangle \\
&= C\langle [e_s \cos(\omega_s t + \varphi_s) + e_L \cos(\omega_L t + \varphi_L)]^2 \rangle \\
&= C\Big\langle \frac{1}{2} e_s^2 [1 + \cos 2(\omega_s t + \varphi_s)] + \frac{1}{2} e_L^2 [1 + \cos 2(\omega_L t + \varphi_L)] \\
&\quad + e_s e_L \cos[(\omega_s - \omega_L)t + \varphi_s - \varphi_L] + e_s e_L \cos[(\omega_s + \omega_L)t + \varphi_s + \varphi_L] \Big\rangle
\end{aligned}
$$

Again, the detector cannot respond to frequencies as high as $2\omega_s$, $2\omega_L$, or $(\omega_s + \omega_L)$. It can, however, respond to $(\omega_s - \omega_L)$ if this is low enough, say, <1 GHz. In this case we shall have a detector current given by

$$
i_d' \sim \frac{1}{2} e_s^2 + \frac{1}{2} e_L^2 + e_s e_L \cos[(\omega_s - \omega_L)t + \varphi_s - \varphi_L]
$$

Since

$$
P_s \sim \frac{1}{2} e_s^2, \; P_L \sim \frac{1}{2} e_L^2
$$

and writing

$$
\omega_s - \omega_L = \omega_{IF}
$$

$$
\varphi_s - \varphi_L = \varphi_{IF}
$$

we can simplify the expression for i_d' to

$$i_d' \sim P_s(t) + P_L + 2[P_s(t)P_L]^{1/2} \cos(\omega_{IF} t + \varphi_{IF}) \tag{5.15}$$

ω_{IF} is called the *intermediate frequency* and will be familiar to readers who have experience with "superhet" radio techniques. We call this *coherent* detection for reasons which will be clearer shortly.

Look carefully at (5.15). It tells us that we have two dc terms (P_s and P_L) followed by a term that represents a frequency which is now in the electronic (as opposed to the optical) range. Furthermore, its amplitude is dependent upon the product of the power levels (P_s, P_L) of the two optical sources.

Consider now the optical-fiber communications system shown in Figure 5.24. Here the "pure" signal source is intensity modulated in the usual way. At the detector it is "mixed" on the photodetector surface with the signal from the other pure source: the local oscillator (LO). With dc filtering in the detector circuit, the only term which remains in the current from the detector is

$$i_d'' \sim 2[P_s(t)P_L]^{1/2} \cos(\omega_{IF} t + \varphi_{IF}) \tag{5.16}$$

This has two advantages. First, it is of narrow bandwidth, fixed by the spectral widths of the two sources ($\omega_{IF} = \omega_s - \omega_L$), of which we will discuss more later. Secondly, the signal amplitude $P_s(t)$ is boosted by the $P_L^{1/2}$ term which effectively provides optical amplification, since the local oscillator contains no signal information, does not pass through the fiber, and we are free to make P_L as large as we like. This optical amplification has the effect of increasing the received power, over the direct detection arrangement, by a factor [see (5.13) and (5.16)] of

$$2\frac{(P_s P_L)^{1/2}}{P_s} = 2\left(\frac{P_L}{P_s}\right)^{1/2}$$

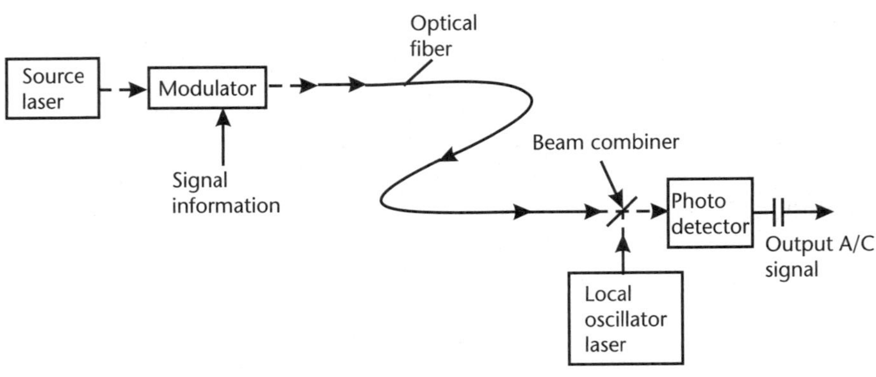

Figure 5.24 Basic coherent optical-fiber communications system.

Since the SNR increases as the square root of the received power, this means that the SNR rises by a factor of

$$\sqrt{2}\left(\frac{P_L}{P_s}\right)^{1/4}$$

Let us take an example. Suppose that both sources have an output power of 1 mW. The signal source suffers attenuation in the fiber of, say, 50 dB, so the optical source power emerging is $10^{-5} \times 10^{-3} = 10^{-8}$ W. This is mixed with 1 mW from the local oscillator source to give amplification by a factor of

$$2\left(\frac{P_L}{P_s}\right)^{1/2} = 2\left(\frac{10^{-3}}{10^{-8}}\right)^{1/2} = 632.5$$

The SNR increases by

$$\sqrt{2}\left(\frac{10^{-3}}{10^{-8}}\right)^{1/4} = 25.14 = 14 \text{ dB}$$

A 14-dB improvement in SNR is well worth having! Up to about 20-dB improvement can be obtained with this technique. This is a valuable advantage since it easily could be equivalent to another 100 km of communications distance.

What is the price to be paid for this improvement?

Fairly clearly, one price is that of the required purity of the lasers used for signal and local oscillator sources. Looking again at (5.15), it is evident that ω_{IF} must have a value which does not stray by more than about 1% of the bandwidth of the modulation signal $P_s(t)$. If it did, it would corrupt and distort $P_s(t)$ and thus effectively introduce noise into the system. Since $\omega_{IF} = \omega_s - \omega_L$, this means, in turn, that neither ω_s nor ω_L can stray by more than about 0.5% of the signal bandwidth. Look yet again at (5.15): φ_{IF} also needs to be stable for the same reason as ω_{IF}, and $\phi_{IF} = \phi_s - \phi_L$, so the phases need to be locked together to the same kind of accuracy.

Let us now insert some numbers into all of this. Suppose that the signal modulation bandwidth is ~ 1 GHz (a 1-Gbs^{-1} digital system for example). Each laser now has to be stable to 0.5% (i.e., to at least 5 MHz), or 1 part in 10^8 of the optical frequency. Single longitudinal mode semiconductor lasers can be fabricated using distributed feedback (DFB) reflectors for mirrors (i.e., mirrors which are Bragg gratings and reflect one very narrow band of wavelengths). Using external distributed Bragg grating reflectors (DBR), laser line-widths as narrow as 10 kHz have been obtained. However, the output frequency and phase drift with temperature in these devices, so they must be loop-stabilized. Figure 5.25(a) shows schematically how this can be done. The source laser is bled by a small amount and compared with an ultra-stable frequency from, for example, an atomic secondary standard, to provide an error signal proportional to the drift. This signal is fed to a temperature or current controller to correct the drift. A similar arrangement is used at the detector end [Figure 5.25(b)], only this time it is the much lower frequency IF which must be maintained constant. A stable electronic oscillator is now required for the reference, but phase error, in addition to frequency error, is required, so as to maintain both ω_{IF} and φ_{IF} constant.

With this kind of stability in the channel frequency it follows that the 1-GHz signal needs only about 10 GHz of channel separation as opposed to 10,000 GHz in the IM/DD case. Hence it becomes possible to run ~ 1,000 channels in each of the two silica windows, thus increasing the bandwidth distance product by a factor ~ 1,000.

Clearly, the requirements on the stability of the lasers are stringent. The lasers need to be not only of stable, narrow line-width, but also tunable, to allow the locking action for frequency and phase. Such lasers are expensive and temperamental, and more work needs to be done before their performances are wholly satisfactory for coherent communications links.

There is one more important problem: (5.15) is valid only if the two signals have *exactly the same polarization*. It is not difficult to arrange that each laser has a stable, linearly polarized output but, as we know, the fiber can play a variety of polarization tricks on light propagating within it. Consequently, the signal light falling on the photodetector is likely to have a varying polarization and hence the IF signal will be subject to fading.

There are two possible types of solution to this problem. The first is to lock the polarization of the signal by means of a feedback loop and polarization controller as in PMD mitigation schemes. The reference polarization in this case can be that of the local oscillator. The second is a "diversity" arrangement (Figure 5.26) where the two orthogonal, linear components are separated (by a Wollaston prism, for example) and IF-detected separately. The two IF signals can then be added electronically. Such systems can be made to work satisfactorily but, again, they add complexity, cost, and unreliability.

The full, reliable implementation of coherent systems with their significant advantages of high channel selectivity (and therefore ease of multiplexing) and of high sensitivity (20 dB) will come. However, as of the time of this writing, the requirement for them has been upstaged by the enormous bandwidths offered by dense wavelength-division multiplexed (DWDM) systems. These latter can deploy up to 100 channels on a single fiber by employing a separate wavelength for each channel, using highly selective Bragg filters.

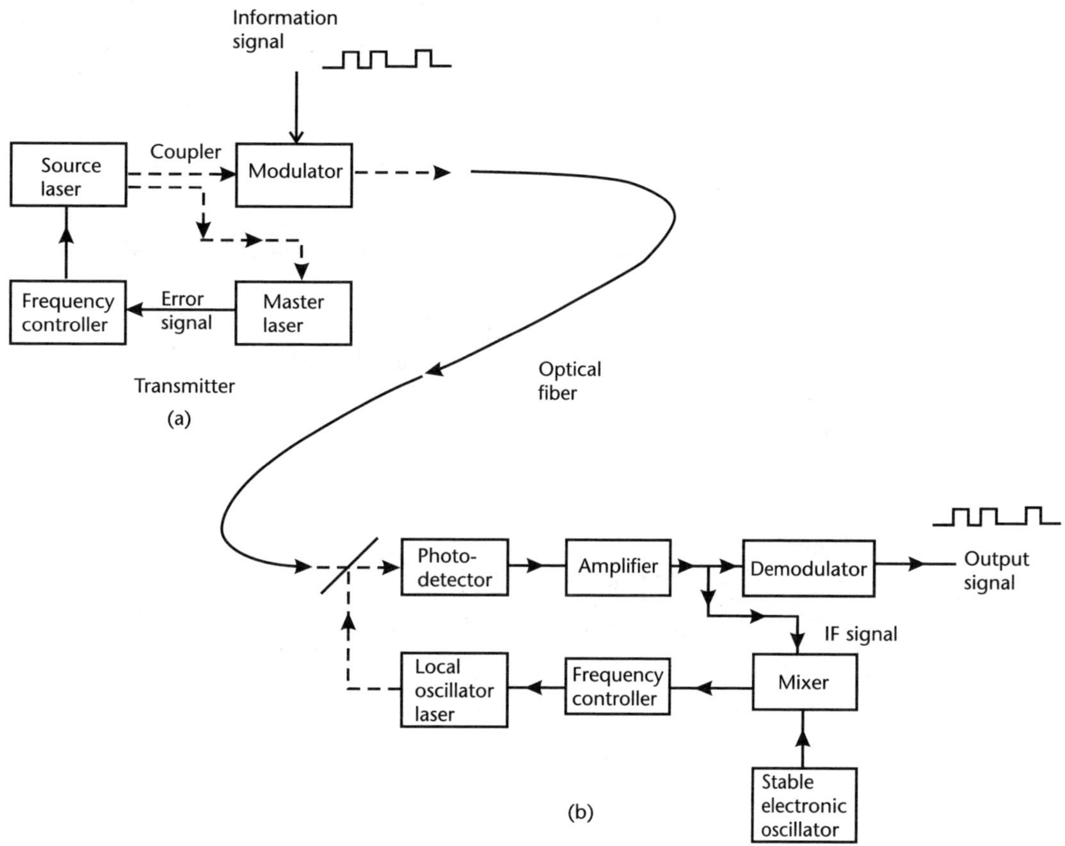

Figure 5.25 Stabilization for a coherent optical-fiber communications system: (a) transmitter; and (b) receiver.

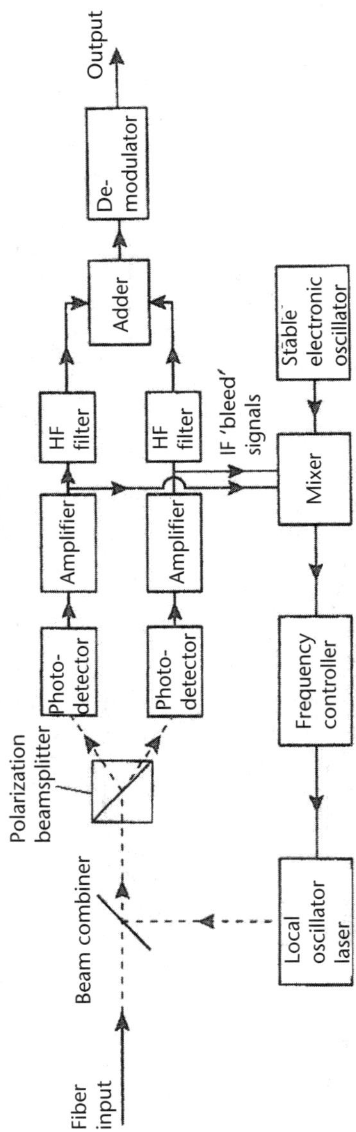

Figure 5.26 A coherent system with polarization diversity.

5.6 Conclusion

In this chapter we have seen how important a part polarization effects can play in high-performance telecommunications systems. It is clearly necessary to understand fully the relevant polarization phenomena in order to both utilize their advantages and reduce their deleterious effects. We have explored the nature of PMD, its description, ways to measure it, and ways to mitigate it. In the next chapter we consider a very different applications area.

References

[1] Poole, C. D., and J. Nagel, "Polarization Effects in Lightwave Systems," in *Optical Fiber Communications III*, I. P. Kaminow and T. L. Koch, (eds.), New York: Academic Press, Vol. B, 1997.

[2] Poole, C. D., et al., "Polarization Dispersion and Principal States in a 147 km Undersea Lightwave Cable," *IEEE JLT*, Vol. 6, 1988, pp. 1185–1190.

[3] Poole, C. D., and R. E. Wagner, "Phenomenological Approach to PMD in Long Single-Mode Fibers," *IEE Elect. Lett.*, Vol. 22, 1986, pp. 1029–1030.

[4] Poole, C. D., and C. R. Giles, "Polarization-Dependent Pulse Compression and Broadening Due to PMD in Dispersion-Shifted Fiber," *Opt. Lett.*, Vol. 13, 1988, pp. 155–157.

[5] Poole, C. D., "Measurement of PMD in Single Mode Fibers with Random Mode Coupling," *Opt. Lett.*, Vol. 14, 1989, pp. 523–525.

[6] Curti, F., et al., "Statistical Treatment of the Evolution of the PSPs in Single-Mode Fibers," *IEEE JLT*, LT-8, 1990 pp. 1162–1166.

[7] Rashleigh, S. C., et al., "Polarization Holding in Birefringent Single-Mode Fibers," *Opt. Lett.*, Vol. 7, 1982, pp. 40–42.

[8] Heffner, B. L., "Automated Measurement of PMD Using Jones Matrix Eigen-Analysis," *IEEE PTL*, Vol. 4, 1992, pp. 1066–1069.

[9] Poole, C. D., and D. L. Favin, "PMD Measurements Based on Transmission Spectra Through a Polarizer," *IEEE JLT*, LT-12, 1994, pp. 917–929.

[10] Rogers, A. J., "Polarization-Optical Time Domain Reflectometry," *IEE Elect. Lett.*, Vol. 16, No. 13, 1980, pp. 489–490.

[11] Shatalin, S. V., and A. J. Rogers, "Location of High PMD Sections of Installed-System Fiber," *IEEE JLT*, Vol. 24, No. 11, 2006, pp. 3875–3881.

[12] Huttner, B., "Local Birefringence Measurements in Single-Mode Fibers with Coherent Frequency-Domain Reflectometry," *IEEE PTL*, Vol. 10, 1998, pp. 1458–1460.

[13] Sunnerud, H., et al., "Technique for Characterization of PMD Accumulation Along Optical Fibers," *IEE Elect. Lett.*, Vol. 34, 1998, pp. 397–398.

[14] Ono, T., et al., "Polarization Control Method for Suppressing PMD Influence in Optical Transmission Systems," *IEEE JLT*, Vol. 12, No. 5, 1994, pp. 891–898.

[15] Noe, R., et al., "PMD at 10, 20, and 40 GBs–1 with Various Optical Equalizers," *IEEE JLT*, Vol. 17, No. 9, 1999, pp. 1602–1616.

[16] Winters, J. H., and M. A. Santoro, "An Electrical Distortion Analyzer for PMD Compensation," *IEEE PTL*, Vol. 2, 1990, pp. 591–596.

6

Polarimetric Optical-Fiber Sensing

6.1 Introduction

The propagation of light along an optical fiber is sensitive to a range of waveguide conditions, geometrical and material. For example, the loss and the polarization state depend upon bends and twists in the fiber; the polarization state depends upon electric and/or magnetic fields to which the fiber might be subjected. The sensitivity of the propagation characteristics to a wide variety of external influences implies that optical fibers could be very valuable in making measurements of external fields, or *measurands*, as they are called in the measurement-sensing context [1, 2].

Clearly, optical-fiber sensors can be easily interfaced with fiber communications or telemetry links for convenient transfer of the information, but this is only one of the advantages offered by such sensors.

First, the fiber is thin and flexible and thus readily installed in a variety of geometrical configurations, either retrospectively or within the original system design. Second, fibers are fabricated from a material which is a dielectric and electrically passive; they require no electrical supply at the point of measurement; neither do they provide electrically conducting paths. They can thus lead to sensors which are intrinsically safe in hazardous environments; this is an attractive feature in, for example, the mining, aeronautical, or petrochemical industries. Third, the one-dimensional nature of the optical fiber allows it to be used to line-integrate or to differentiate the measurand field along its length. The advantages of these features will become more apparent when we consider some examples of their use.

Clearly, there are several special problems which must be considered when designing sensing systems as opposed to communications systems. One obvious problem immediately arises: Since the fibers are sensitive to so many external influences, how do we ensure that a given fiber is sensitive to only one wanted parameter at a time? And a related question: Is this large range of sensitivities compatible with the much vaunted interference-immunity of optical-fiber telecommunications? The first question is crucial for the design of optical-fiber sensing systems. To answer it for any given measurement function means that the designer must use considerable optoelectronic engineering skill and ingenuity. He or she must also rely heavily on a basic understanding of behavior and on the ability of the technologist to provide fiber with carefully controlled, and predefined, properties.

The answer to the second question is more straightforward. To avoid any sensitivity to external influences in the telecommunications fiber, one must ensure that the optical detection system is not sensitive to any changes effected by external fields: this implies either protection from them or, in the case of polarization effects, a detection system which is not sensitive to the light's polarization state, but only to the total received power. With care, this can be arranged.

The skill and ingenuity of the sensor system designer can be assisted considerably by the use of specialized fibers. In addition to the basic fiber properties already described, special properties, of use both to telecommunications and to sensing technologies, can be added to the fibers. One of the processes by which this is done is the doping of the fiber with, for example, elements such as neodymium, erbium, praseodymium, yttrium, or germanium. These elements provide, variously, temperature-dependent absorption, fluorescence spectra, or enhanced Raman-scatter coefficients. (The fluorescence properties of the rare-earth dopant erbium are famously useful in fiber lasers and fiber amplifiers [3].)

Finally, there are fibers with special coatings. Telecommunications fibers need coatings in order to make them rugged. A soft primary coating is put onto the fiber as it is drawn from the preform melt, in order to protect it from atmospheric attack (especially from moisture) while its surface is bare and vulnerable. A much harder, secondary coating is added subsequently to provide strength for handling and duct installation. A sensor fiber also needs to be made rugged, but, additionally, it needs a coating which will enable it to survive in the measurement environment, and to interact optimally and consistently with the measurand field. Some special sensor coatings are already available: metal coatings, polyimide coatings, and carbon surface-impregnation, for example. However, very much more needs to be done in this area if fiber sensing systems are to be matched properly to their measurand environments.

In general terms, the phase of a lightwave propagating in an optical fiber is more sensitive to external influences than any other propagation parameter, and thus, optical-fiber sensors which utilize this phase dependence are capable of the highest measurement sensitivities.

In this class of sensors we find those which rely on the disturbance of absolute phase (normally called interferometric sensors), as well as those which rely on the relative phase displacement of two polarization eigenmodes (which are called polarimetric sensors).

As each allowable propagation mode in a fiber possesses its own definite phase and polarization properties, it is only with monomode fiber that single, definable phase and polarization states exist at each point along the fiber (though these may vary from point to point), and thus it is only with monomode fiber that one may expect optimally to extract, from either kind of state, information imposed on the propagating light by fields which one is seeking to measure.

Both polarimetric and interferometric sensors in this class are essentially forms of Mach-Zehnder device, that is to say, optical phase "bridges" [Figure 6.1(a)] allowing detection of the phase disturbance (produced by the measurand) by comparison of phases in two "arms," which may or may not be within a single fiber. In polarimetric devices [Figure 6.1(b)] the two arms are constituted from the two polarization eigenmodes of a single monomode fiber, and the phase differential introduced by the measurand is inferred from the resultant (compound), changed, polarization state.

The one-dimensional nature of the optical fiber makes it especially valuable for sensing measurands which are the line integrals of imposed fields, and also

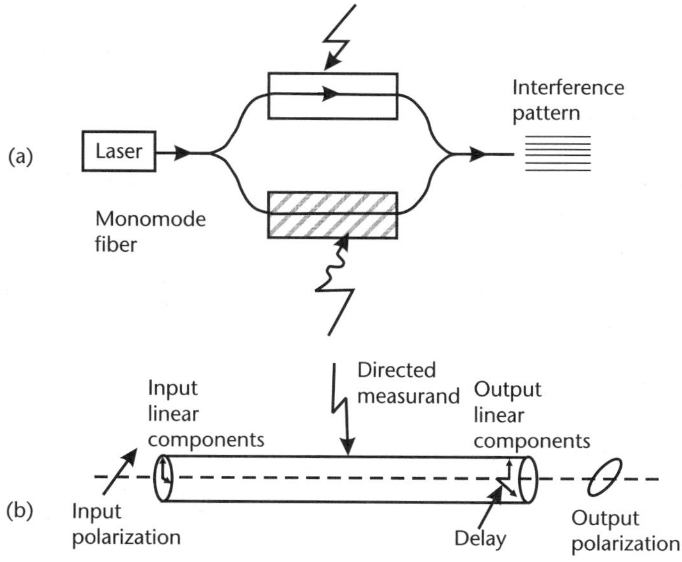

Figure 6.1 (a, b) The Mach-Zehnder principle.

for measuring the spatial distributions of fields, over any chosen path, using techniques of line differentiation. Polarimetric techniques are particularly well suited to these methods, as will be illustrated below.

The primary problem for polarimetric optical-fiber sensors (as with most sensors) is that of maximizing the signal-to-noise ratio, which means, specifically, arranging for maximum phase differential per unit of measurand field and, simultaneously, minimum disturbance of the polarization state by unwanted influences (noise).

These various features have led to the whole new discipline of optical-fiber sensing, but this chapter will deal just with that branch of this discipline which uses polarimetric effects.

Three fairly distinct categories of optical-fiber sensor have evolved: (1) those which make measurements at a single point (point sensors); (2) those which make measurements at a set of discrete, predetermined points (quasi-distributed or, sometimes, multiplexed sensors); and (3) those which can make measurements anywhere along the length of a fiber with some specified spatial resolution (distributed, or fully distributed, sensors). In addition there might be added the category which either averages the measurand field over a long length of fiber (averaging sensors) or line-integrates the field over the fiber length (integrating sensors). We shall give examples of most of these categories within the limited framework of polarization phenomena in optical fibers. In this chapter we deal with point and fully distributed polarimetric sensors. Quasi-distributed sensors are left until the next chapter because they (almost exclusively) rely on nonlinear polarimetric effects.

6.2 Point Sensors

Optical-fiber sensors which address the value of a measurand field at a particular point have the advantage, over conventional technology, of a readily variable sensitivity, via the length of fiber used. In order to limit the spatial extent of the measurement, the fiber can be coiled or spiraled, if required.

Some examples of interferometric and polarimetric optical-fiber point sensors will now be given.

6.2.1 Interferometric Sensors

Two light beams will interfere optimally, that is to say, to produce an interference pattern of maximum visibility, when they are mutually coherent and have the same polarization state. If a phase disturbance is inserted in the path of one of two such beams, the interference pattern will shift, and measurement of this shift is the basis for several sensing methods.

The Mach-Zehnder Interferometer

Figure 6.2 encapsulates the essence of an interferometric optical-fiber sensor. It is known as a Mach-Zehnder interferometer and comprises two monomode fiber paths: one of which is exposed to the measurand (a pressure or a temperature, for example), and the other of which is isolated from it. The effect of the measurand is to alter the refractive index of the fiber and thus the phase of the light in the exposed arm, and hence to vary the output level, which results from the optical interference between the signals from the two arms.

Let us suppose that equal light powers are launched into each of the two arms, and with the same polarization state. On emergence at the output we may write the sum of the two interfering electric field amplitudes as

$$E_T = E \sin \omega t + E \sin(\omega t + \varphi)$$

where φ is the phase disturbance introduced by the action of the measurand.

The received (by the photodetector) light intensity will be proportional to the square of the modulus of the resultant field (Section 1.2.2); that is,

$$|E_T|^2 = E^2(1 + \cos \phi)$$

or

$$I_T = I(1 + \cos \varphi)$$

where I is the light intensity.

Now because

$$dI_T = -I \sin \varphi \, d\varphi$$

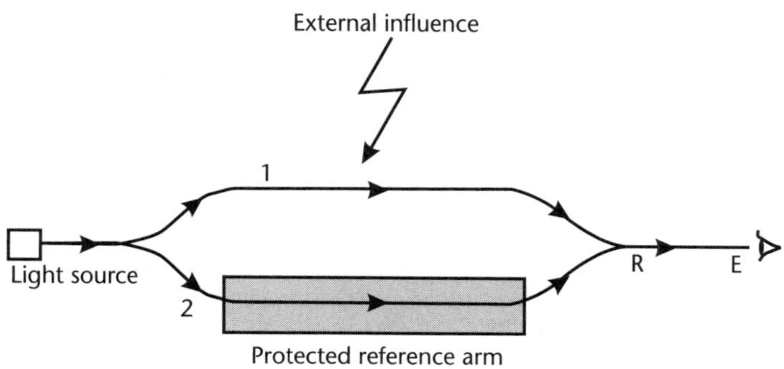

Figure 6.2 An optical fiber Mach-Zehnder interferometer.

the variation in I_T with changes in φ will be greatest when it is biased at the value of $\pi/2$. This is known as the quadrature condition and is a necessary complication for good operation of a Mach-Zehnder device.

It also is clear that the coherence length of the source must exceed any difference in optical path introduced by the measurand (for maximum visibility of the interference pattern) and that, of course, the two beams should emerge still with the same polarization state. Any deviation from either of these two conditions will reduce the sensitivity of the measurement.

The Mach-Zehnder arrangement has a significant advantage in that there is independence from any common-mode effects on the lead fibers (source to M-Z, M-Z to detector) since these effects will not alter the relationship between the two signals.

A special example of a Mach-Zehnder arrangement, known as a Michelson interferometer, is shown schematically in Figure 6.3. Polarized light is launched into a fiber and is partially reflected at the far end of the fiber; this acts as a reference beam. The rest of the light passes out of the fiber and is reflected from a mirror attached to an object, whose position is to be detected, and then passes back into the fiber. The two returning components are then bled off by an optical coupler to interfere on the surface of a photodiode. Clearly, the displacement of the reflective object will affect the relative phase of the two components, and thus the interference signal, which can be used to determine the change in the position of the object to nanometer precision.

The Optical-Fiber Gyroscope

A very special form of Mach-Zehnder interferometer is shown in Figure 6.4. This is the optical-fiber gyroscope, and is one of the most highly developed optical-fiber sensors in existence [4].

The two phase arms now are within the same loop of fiber, but are distinguishable by virtue of the fact that they propagate in opposite directions around the loop. On emergence from their respective ends, the two signals are allowed to interfere. If the loop rotates about an axis through its center and normal to its plane, one of the signals advances in phase (against the rotation) while the other (with the rotation) is retarded. The resultant perturbation of

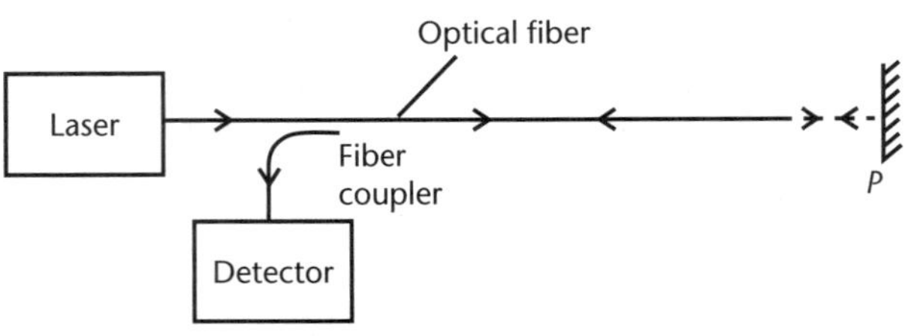

Figure 6.3 An interferometric displacement sensor.

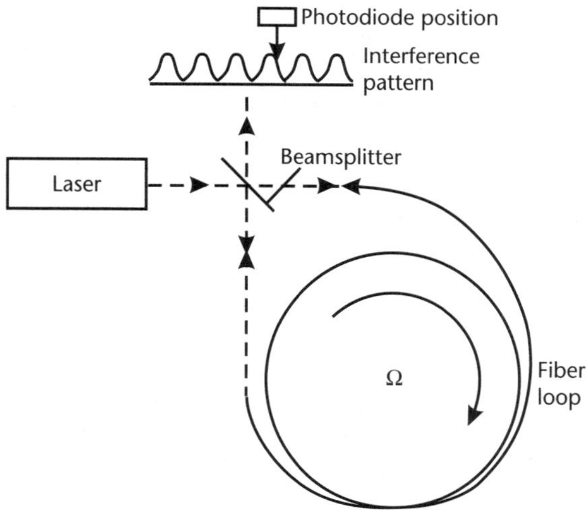

Figure 6.4 The optical-fiber gyroscope.

the interference pattern allows the magnitude of the rotational velocity to be determined. (The perturbational effect is known as the SAGNAC effect after the Frenchman who discovered it, in 1913.)

The optical-fiber gyroscope epitomizes a central problem in most optical-fiber sensor technology, and thus, it is instructive to study it in more detail. In order to establish the technique as a viable operational measurement method, it must compete directly with the existing conventional techniques, which are well tried and understood by the practitioners. To do this it must offer at least one singular advantage and perform at least as well in all other respects, including cost.

Gyroscopes are very important devices for navigation and automatic flight control. The conventional gyroscope based on the conservation of angular momentum in a spinning metal disc is highly developed, but contains parts which take time to be set in motion ("spin-up" time) and which wear. The device is also relatively expensive both to install and to maintain.

The optical-fiber gyroscope overcomes all these problems (but, inevitably, has some of its own). Consider the arrangement shown in Figure 6.4. Light from a laser is fed simultaneously into the two ends of a fiber loop, via the beamsplitter, so that two beams pass through the loop, in opposite directions. When the beams emerge at their respective ends, they are brought together, again via the beamsplitter, and interfere on a receiving screen. This arrangement can be regarded as a special form of Mach-Zehnder interferometer, where the two arms of the interferometer lie within the same fiber, but the two signals traverse it in opposite directions. Clearly, under these conditions a sinusoidal interference pattern will be formed on the screen and, if the two beams are equal in intensity, the visibility will be 100%.

Suppose now that the complete system is rotating clockwise at an angular velocity Ω. In this case the clockwise propagating beam will view the end of the fiber receding from it as it travels, and it will thus have farther to go before it can emerge. Conversely, the anticlockwise rotating light will see its corresponding end approaching, and will have less far to go. The consequence is thus a relative phase shift between the two beams and a consequent shift in the interference pattern on the screen. (This is, in fact, a somewhat simplistic explanation of the physics involved. A rigorous explanation requires the help of the general theory of relativity, since rotating systems are accelerating systems, but the explanation given here is correct to the first order.) It follows that the change in the interference pattern can be used to measure the rotation, Ω, by placing a photodiode (for example) at a position on the screen where it can record a linear variation of received power with lateral shift (Figure 6.4).

The phase shift caused by the angular rotation is readily calculated (see Figure 6.5). Suppose that there are N turns of fiber on the coil and that the coil radius is R. Then, in the absence of rotation, the time of flight around the coil will be given by

$$\tau = \frac{2\pi RNn}{c_0} \tag{6.1}$$

where c_0 is the velocity of light in free space and n is the refractive index of the fiber material. If the coil is rotated about an axis through its center and perpendicular to its plane, with angular rotation Ω, then the fiber ends will have rotated through an angle $\Omega\tau$ while the light is propagating in the fiber, and thus through a distance $\Omega\tau R$. Hence the difference in distance traveled by the two counter-propagating beams will be twice this; that is,

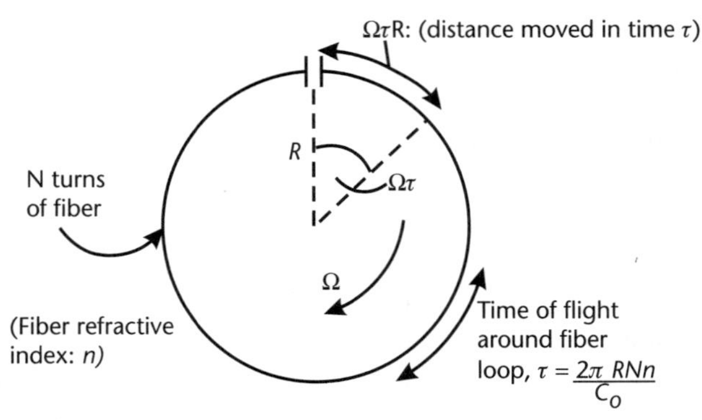

Figure 6.5 Gyroscope geometry.

$$dl = 2\Omega \tau R \tag{6.2}$$

Now the clockwise and anti-clockwise propagating light components no longer propagate with the same velocity when the coil is rotating, as viewed from the original stationary ("inertial") frame. We have to take account of the so-called Fresnel drag, whereby light propagating in a medium which is moving with velocity v in the same direction as the light will propagate, relative to the stationary frame with a velocity given by (see, for example, [4])

$$c_v = \frac{c_0}{n} + v\left(1 - \frac{1}{n^2}\right)$$

where $(1 - n^{-2})$ is called the Fresnel-Fizeau drag coefficient. (This is a direct consequence of special relativity.) Hence, in our case the two velocities around the loop are given by

$$c_+ = \frac{c_0}{n} + R\Omega\left(1 - \frac{1}{n^2}\right) \tag{6.3}$$

$$c_- = \frac{c_0}{n} - R\Omega\left(1 - \frac{1}{n^2}\right)$$

Consider now the difference between the times of arrival at the end of the fiber coil for the two counterpropagations. We have

$$t = \frac{l}{v};\ dt = \frac{dl}{v} - \frac{l}{v^2}\,dv \tag{6.4}$$

and

$$dl = 2\Omega\tau R = \frac{4\pi R^2 N n \Omega}{c_0} \qquad \text{(from (6.1))}$$

$$v = \frac{c_0}{n};\ l = 2\pi RN \qquad \text{(from (6.3))}$$

$$dv = 2R\Omega\left(1 - \frac{1}{n^2}\right)$$

Substituting the latter expressions into (6.4), we have

$$dt = \frac{4\pi R^2 N\Omega}{c_0^2} \tag{6.5}$$

Note that this is *independent of the refractive index n* and therefore is independent of the fiber medium. (This is a common source of confusion in regard to the operation of the optical-fiber gyroscope.) The phase difference between the two counter-propagations when the coil is rotating is now easily constructed from (6.5) as

$$\Phi = \omega\, dt = \frac{2\pi c_0}{\lambda_0}\, dt = \frac{8\pi^2 R^2 N\Omega}{c_0 \lambda_0}$$

where λ_0 is the free space wavelength. This can also be written:

$$\Phi = \frac{8\pi A\Omega}{c_0 \lambda_0} \tag{6.6a}$$

or

$$\Phi = \frac{2\pi LD\Omega}{c_0 \lambda_0} \tag{6.6b}$$

where A is the total effective area of the coil (i.e., the total area enclosed by N turns), L is the total length of the fiber, and D is the diameter of the coil.

Let us now insert some numbers into (6.6b). Suppose that we use a wavelength of 1 μm with a coil of length 1 km and a diameter of 0.1m. This gives

$$\Phi = 2.1\Omega$$

For the Earth's rotation of 15° h^{-1} (7.3×10^{-5} radians s^{-1}) we must therefore be able to measure ~ 1.5×10^{-4} radian of phase shift. This can quite readily be done. In fact, it is possible, using this device, to measure ~ 10^{-6} radian of phase shift, corresponding to ~ 5×10^{-7} radians s^{-1} of rotation rate.

What, then, are the problems? First, since the fringe visibility will only be 100% if the two interfering beams have the same polarization, it is necessary to use polarization-maintaining monomode fiber: usually linearly birefringent (hi-bi) fiber is used. Secondly, there is a problem with the polarization-optical Kerr effect. The electric field of one beam will act, via the optical Kerr effect, to alter the phase of the other. The effect is small but then so is the phase difference which we are seeking to measure, at low rotation rates. This effect we can calculate but, in order to do this, we need some ideas from nonlinear optics, to be covered in the next chapter: the relevant calculation will be performed there (Section 7.2).

Other sources of noise are: Rayleigh backscatter in the fiber, which gives a coherent interfering signal, and drift in the value of the area of the fiber, due to temperature variation.

All these problems make it difficult for this device to compete successfully at the lowest rotation rates, although its simplicity and relatively low cost give it a distinct advantage for rotation rates ~ $0.1°\ h^{-1}$ and above. To achieve the highest sensitivity, the so-called minimum-configuration design is employed. This is shown in Figure 6.6. It ensures that the common, reciprocal path comprises almost the entire system. Integrated optics is used to assist in this.

Polarization of the light and the beam-splitting are performed on the integrated-optical (I/O) chip. Also on the chip are a frequency shifter (acousto-optical) to allow the direction of rotation to be determined, and a phase modulator to ensure that the detection bias is maintained at its point of maximum sensitivity. The I/O chip also means that the device can be very compact; one has been built small enough to be enclosed in a sardine tin!

(I/O chips, such as this, which comprise both optical and electronic functions, are sometimes referred to as optoelectronic integrated-optical circuits, or OEICs.)

Applications of the minimum configuration optical-fiber gyroscope range from ballistic missiles, through the location and control of oil-well drill tips, to motor vehicle navigation systems.

The optical-fiber gyroscope is in direct competition with the ring laser gyroscope (RLG), shown diagrammatically in Figure 6.7, for many applications. This device uses the same Sagnac principle but does not use optical fiber; rather, it uses a triangular laser cavity, cut in a material such as quartz, which is filled

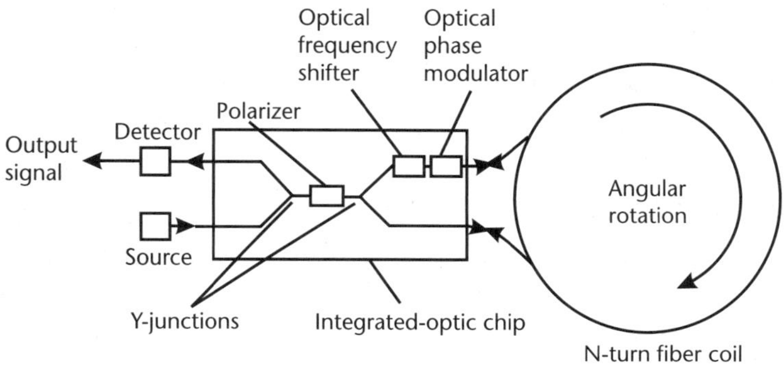

Figure 6.6 "Minimum configuration" fiber gyroscope.

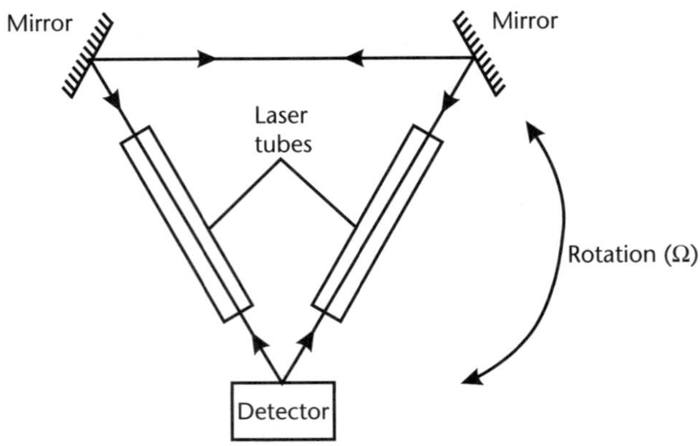

Figure 6.7 The ring laser gyroscope.

with a laser gain medium (e.g., He-Ne). The RLG provides a difference frequency between the two counterpropagating laser modes, a difference resulting from the Sagnac effect—it is proportional to the rotation rate.

This RLG device is more highly developed at the present time than the optical-fiber gyroscope, although it still has its problems. The fiber gyroscope has the potential for equivalent performance and also for being a much cheaper alternative to the RLG.

6.3 Line-Integrating Polarimetric Sensors

The (quasi-) one-dimensional nature of the optical fiber implies that it traces a line within the measurand field. This allows us, when convenient, to perform a line integration of the field. For example, the line integral of the magnetic field around a current-carrying conductor is equal to the value of the current flowing in the conductor, from Ampere's circuital theorem; the line integral of the electric field between two points is equal to the voltage between the points. Hence, both current and voltage can, in principle, conveniently be measured in this way, and we shall in this section describe devices capable of making these measurements.

6.3.1 Optical-Fiber Current Measurement

A prevalent problem in the electricity supply industry (ESI) throughout the world is that of the measurement of the electric current produced by generating stations of all types. The current generated is distributed to customers of the supply company on a grid of high voltage lines (since the higher the voltage the smaller is the current for a given transfer of power and thus the smaller are the resistive losses). This current must be measured accurately. This measurement is necessary to ensure that the customer is charged the correct amount for the energy that is used; but it is also necessary for proper control of the grid network, and for rapid indication of fault conditions, to allow corrective action.

The current usually is measured at distribution junctions known as switching substations, and it must be done on high voltage bus-bars, the voltage level being several hundred kilovolts.

Conventional current transformers, using primary and secondary induction windings, are fairly widespread for this measurement purpose but they require costly high voltage insulation between the two windings, and they inevitably suffer from saturation effects, owing to ferromagnetic hysteresis, and from relatively low bandwidth, owing to the large winding inductances.

Consider the alternative, optoelectronic approach to high-voltage current measurement illustrated in Figure 6.8. A length of monomode optical fiber (a good insulator) is coiled several times around a current-carrying bar. Linearly polarized light from a semiconductor laser is launched into this fiber coil. While propagating through the coil of fiber, the light comes under the influence of the magnetic field due to the current in the bar. The field lines around such a current are cylindrical in form, so they lie parallel with the fiber axis as it coils around the bar. Hence the linearly polarized light comes under the action of a longitudinal magnetic field as it propagates; it thus experiences the Faraday magneto-optic effect, and the direction of polarization is rotated by an amount proportional to the magnetic field, and to the path length along which the field acts. This effectively is a magnetic-field-induced circular birefringence (see Section 4.2.7).

Now the total rotation of the polarization direction on emergence will be proportional to the line integral of the field around the loop; that is,

$$\rho = NV \oint H \cdot dl \tag{6.7}$$

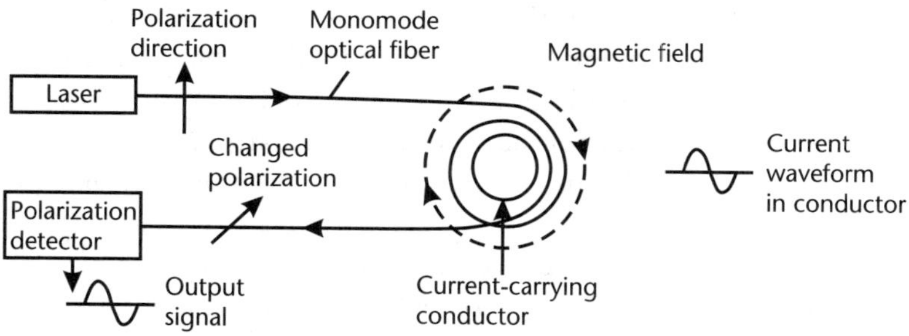

Figure 6.8 Optical-fiber current measurement.

where V is the Verdet (magneto-optic) constant, N is the number of turns on the fiber coil, H is the magnetic field, and 1 is the path length.

But from our knowledge of basic electromagnetism we know that the line integral of the magnetic field around a current-carrying conductor is just equal to the current (Ampere's circuital theorem). Hence (6.7) can be written:

$$\rho = NVI$$

where I is the current to be measured.

Thus, a measurement of ρ allows us to measure I, since both N and V are known constants. This will be a very convenient means for measuring current at high voltage because:

1. The fiber is made from a dielectric, insulating material and so costly insulation will be required between the high voltage bar and the (earthy) indication point.
2. There is no hysteresis effect, since no ferromagnetic materials are involved.
3. The magneto-optic effect is very fast (almost instantaneous) in fused silica, and thus the measurement bandwidth can be very large.
4. The fiber is easily coiled around a bar, and thus installation is very straightforward. How, then, can we actually measure ρ? Suppose that the emerging linearly polarized light falls on to a linear polarizer that is set with its polarization direction parallel with that of the light's *input* polarization direction [Figure 6.9(a)]. In the absence of a magnetic field ($\rho = 0$) all the light will be passed by the polarizer (ignoring its intrinsic attenuation).

Let us assume that the electric field amplitude of the propagating light is E_0, so that an intensity proportional to E_0^2 is passed, in the absence of current. When current flows, the polarization is rotated through an angle ρ and only a field component $E_0 \cos \rho$ will now be passed by the polarizer, giving a measurable intensity proportional to $E_0^2 \cos^2 \rho$. This intensity, in principle, allows ρ to be deduced.

However, there is a more convenient way to measure ρ.

Suppose that instead of a simple polarizer, we use a Wollaston prism, with its polarization axes set at ±45° to the input polarization direction. We now have two intensity outputs from the Wollaston prism [see Figure 6.9(b)]:

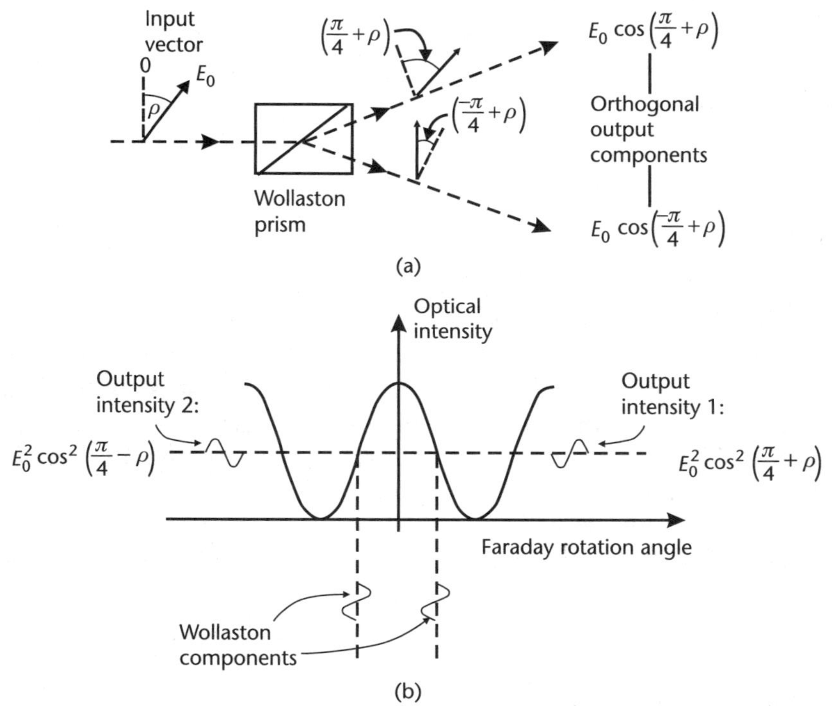

Figure 6.9 Detection actions for current-measurement signals: (a) Wollaston E-field components; and (b) output intensities for Wollaston components.

$$I_1 = KE_0^2 \cos^2\left(\frac{1}{4}\pi - \rho\right)$$

$$I_2 = KE_0^2 \cos^2\left(\frac{1}{4}\pi + \rho\right)$$

where K is the usual universal constant.

If we detect these two intensities separately (by measuring the optical powers falling on two separate photodiodes—remember that power = intensity × area), then we can readily arrange for the electronics to construct the function:

$$S = \frac{I_1 - I_2}{I_1 + I_2} = \frac{\cos^2\left(\frac{1}{4}\pi - \rho\right) - \cos^2\left(\frac{1}{4}\pi + \rho\right)}{\cos^2\left(\frac{1}{4}\pi - \rho\right) + \cos^2\left(\frac{1}{4}\pi + \rho\right)} \tag{6.8}$$

which gives, on manipulation of the functions:

$$S = \sin 2\rho$$

and, if 2ρ is small ($\ll \pi/2$):

$$S \sim 2\rho = 2NVI$$

Hence, S is proportional to the current I, under these conditions, and independent of the light intensity ($\sim E_0^2$), which can vary with time.

What then are the problems with this method for measuring current? The first is that the fiber must be bent around the bar. If a fiber is bent, linear birefringence is introduced since the fiber is strained asymmetrically (Figure 6.10) and strain alters the refractive index via the strain-optic effect. Now we know that, in a linearly birefringent fiber, only the two linear eigenmodes (i.e., those linearly polarized states which lie parallel to the birefringence axes) propagate without change of form. It is clear that a rotating linear polarization state cannot remain linearly polarized within such a fiber. It will emerge in an elliptical state that will depend not only on the bend birefringence but also on the way in which it has rotated along the fiber (i.e., on the current to be measured).

Now (6.8) indicates that the value of S is proportional to the difference between the two Wollaston outputs (I_1, I_2). However, this difference will also depend upon the axes of the polarization ellipse (i.e., on the linear bend birefringence), in addition to the circular birefringence due to the current to be measured. If, for example, the polarization ellipse happens to degenerate into a circle, S becomes zero even though the current is nonzero. Hence, this bend birefringence effect clearly will interfere with the measurement.

The effect can be quantified very conveniently with the aid of the Jones matrices discussed in Section 3.11. The matrix for a polarization element which rotates the polarization through an angle ρ (i.e., which possesses circular birefringence 2ρ) in the presence of a linear birefringence δ, is given by (see Section 3.11.4)

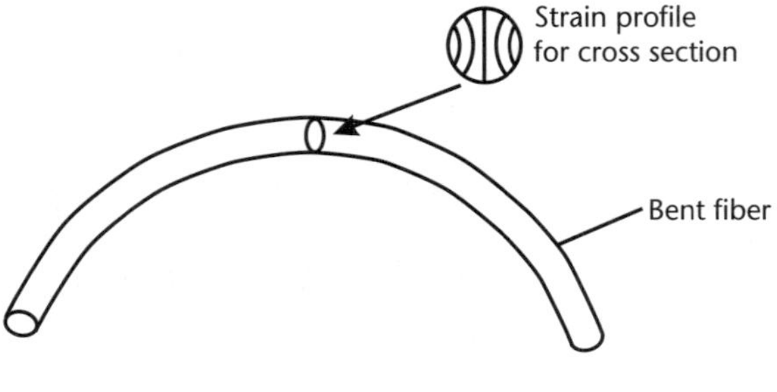

Figure 6.10 Fiber bending-strain profile.

$$M = \begin{pmatrix} \alpha + i\beta & -\gamma \\ \gamma & \alpha - i\beta \end{pmatrix} \tag{6.9}$$

where, with $\Delta = (\rho^2 + \delta^2/4)^{1/2}$ we have:

$$\alpha = \cos \Delta; \qquad \beta = \frac{1}{2}\delta\left(\frac{\sin \Delta}{\Delta}\right); \qquad \gamma = \rho\frac{\sin \Delta}{\Delta}$$

Here, the matrix axes (Ox, Oy) are taken to be those of the linear birefringence fast and slow axes. This matrix is fairly complicated because the interaction between ρ and δ is quite subtle. The effect that δ has on the polarization state depends upon the state itself. For example it has no effect on a linear state parallel with one of the axes, but if that state is rotated through 45°, it has a maximum effect; with the orientation changing continuously as a result of the rotation, ρ, it follows that the effect of δ also is changing continuously.

Suppose, then, that the linearly polarized input light to the fiber coil is launched with its polarization direction aligned with one of the birefringence axes, say, Ox (fast). This can be represented by the column vector:

$$\begin{pmatrix} E \\ 0 \end{pmatrix}$$

where E is its electric field. The polarization state of the output light can now be determined as

$$\begin{pmatrix} E_x \\ E_y \end{pmatrix} = M\begin{pmatrix} E \\ 0 \end{pmatrix}$$

using M from (6.9) (remembering that the Es are all complex numbers). This allows S, from (6.8), to be constructed as

$$S = 2\alpha\gamma = 2\rho\frac{\sin 2\Delta}{2\Delta} \tag{6.10}$$

which quantifies our measurement problem. For, suppose, first, that the bend-induced linear birefringence, δ, is very much larger than the current-induced circular birefringence 2ρ (and thus also very much greater than the rotation ρ). In this case

$$\Delta = \left(\rho^2 + \frac{1}{4}\delta^2\right)^{1/2} \approx \frac{1}{2}\delta$$

and thus from (6.10),

$$S \approx 2\rho\left(\frac{\sin\,\delta}{\delta}\right)$$

Hence, for $\delta > 0$, we see that $S < 2\rho$ and the measurement sensitivity is reduced. In fact, the situation is worse than a simple reduction of output, for the value of δ will be temperature dependent (via the temperature dependence of the strain-optic coefficient) and so S will be temperature dependent. Furthermore, δ can also be induced by vibration (since vibrational pressure also will strain the fiber), and hence, S will suffer from vibrationally induced A/C noise.

What can be done about these problems?

First, the bend birefringence can be kept small by ensuring that the bend diameter is kept large (>0.5m). Second, it is possible to use optical fiber which has an intrinsic circular birefringence (as has an optically active crystal such as quartz, for example). Circular birefringence can be induced in an optical fiber by several methods, one of which is by simply twisting the fiber about its own axis (Section 4.2.4). In this latter case we now have

$$2\rho \gg \delta$$

and hence,

$$\Delta \sim \rho$$

and

$$S \sim \sin\,2\rho$$

(Physically, what has happened here is that the large value of intrinsic birefringence has caused the polarization state to rotate very rapidly along the fiber, thus averaging out the effects of the linear birefringence). However, ρ is now the sum of the intrinsic circular birefringence ($2\rho_0$) and the current-induced circular birefringence ($2\rho_I$), so that

$$S \sim \sin(2\rho_0 + 2\rho_I)$$

But from the discussion in Section 4.2.7 we know that there is a fundamental difference between these two components of circular birefringence: ρ_0 is reciprocal, while ρ_I is nonreciprocal. This means that if the light is back-reflected down the fiber, so that it performs a go-and-return passage through the coil around the conductor, the intrinsic reciprocal birefringence ($2\rho_0$) will be cancelled, while the current-induced, nonreciprocal birefringence ($2\rho_I$) will be doubled. Hence, on back-reflection in an arrangement such as that shown in Figure 6.11, we shall have

$$S \sim 4\rho_I$$

This has the added advantage of removing the temperature dependence of ρ_0 (which is due, again, to the temperature dependence of the strain-optic coefficient when ρ_0 is obtained from twist strain). Yet another advantage of this arrangement is that it has the convenience, for installation, of being single-ended (Figure 6.11).

It is, perhaps, not obvious that, under these conditions, the (cancelled) intrinsic circular birefringence can still be effective in swamping the linear birefringence. That this is indeed the case can best be appreciated by reference, again, to the Poincaré sphere (Figure 6.12). The eigenmodes for both forward and backward propagations are close to the circularly polarized poles. Hence, the action in each direction of propagation (equivalent to rotations about the eigenmode diameters) maintains the state of polarization close to the equator (i.e., close to the linear state). In other words, it is the ratio $2\rho/\delta$ which controls the action in each direction, rather than the resultant value of either quantity after the double passage.

Hence, an alternative approach to the solution of the problem of linear-birefringence interference in this device is to use a fiber with a very large intrinsic

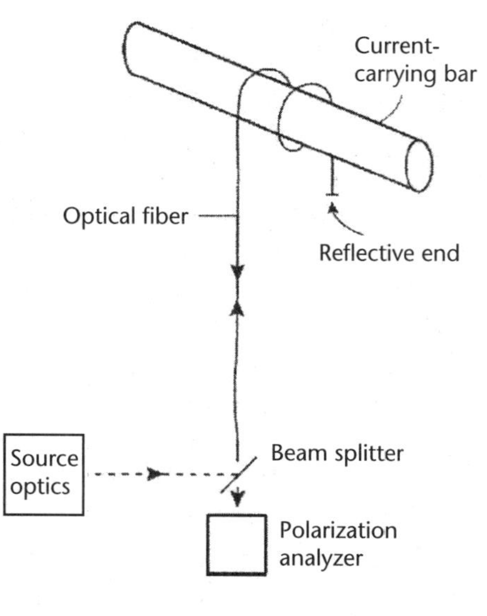

Figure 6.11 Single-ended configuration for current measurement.

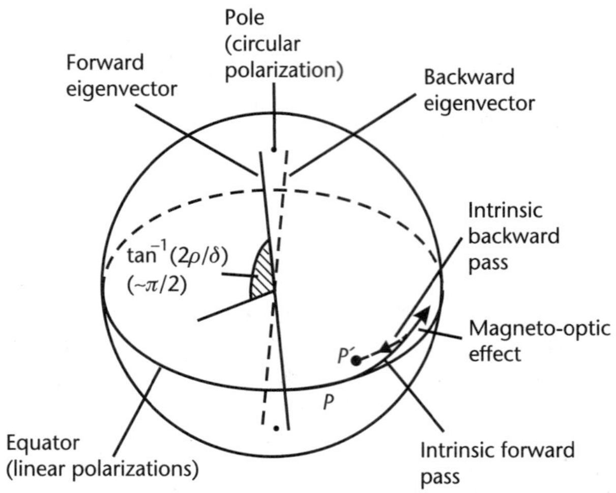

Figure 6.12 Effect of large intrinsic circular birefringence.

circular birefringence. Such a fiber cannot yet (as far as the author is aware) be manufactured, but it is clear from (6.10) that, if 2ρ can always be maintained at a much greater value than δ, then δ can be ignored. (Some circular birefringence may be introduced by twisting the fiber, but the amount is obviously strictly limited by the necessity to avoid physical damage to the fiber.) An obvious difficulty in this case is the possibility that the drift inherent in a large "bias" value of ρ will itself give rise to a significant noise level.

Hence, we can see that our detailed knowledge of polarization optics has allowed us to design a very satisfactory device, free from temperature and vibration effects, which is capable of making a very important measurement, cheaply and conveniently. Devices based on these principles have been used in the electricity supply industry, in various diagnostic and testing procedures where quick, easily installed devices have great advantages.

Figure 6.13 shows a particularly interesting application of optical-fiber current measurement, where it is not even necessary to twist the fiber since the intrinsic circular birefringence is not needed.

A fiber encloses a high-voltage transmission tower, and measures the current flowing into the ground when a short-circuit fault is struck between one of the high-voltage phase conductors and the earthed tower. This measurement is able to provide valuable information on the earth current which flows when such a fault occurs as a result, for example, of a direct lightning strike on the line. This current would be difficult to measure in any other way, and virtually impossible to measure using conventional current transformers. In this application the bend birefringence is not a problem because the coil diameter is very large (~ 10m); the vibration is not a problem because the measurement has

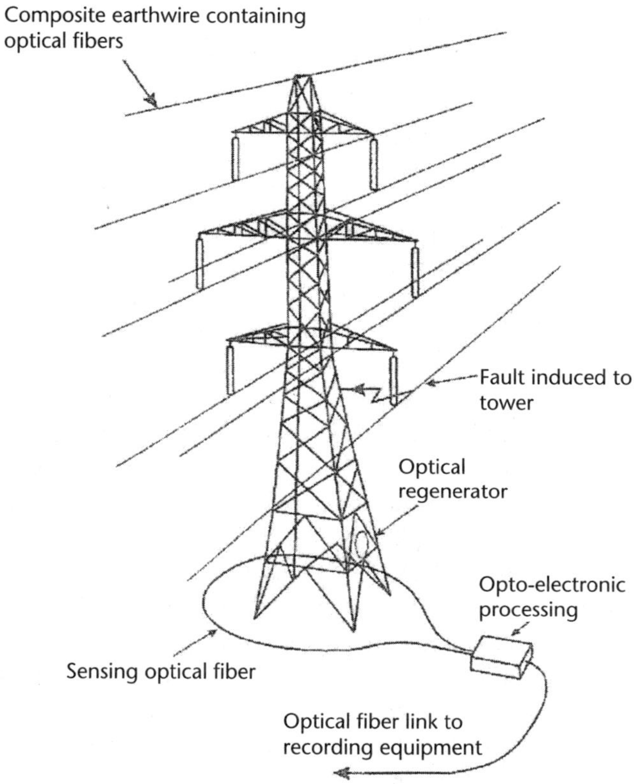

Figure 6.13 "Tower footing" optical-fiber current measurement.

been completed by the time that the mechanical stock of the fault propagates down the tower to the ground; the temperature dependence is not a problem because the temperature drift over such a short time is negligible. On the other hand, the advantages are that the bandwidth is large enough to ensure that the short period (~ 1 μs) waveform is accurately reproduced; and the fiber is installed, and removed, in minutes. This is thus a good example of how the performance of an optoelectronic (or any other) system or device can be matched to the specific requirements with great advantage.

One final note on the bandwidth available with the optical-fiber current measurement device: the speed of the magneto-optic effect is not the limitation, but the bandwidth is limited by the time taken for the light to pass around the fiber loop. Clearly the measurement cannot take place in a time less than this, for the full rotation will not then have occurred. So perhaps just one loop is optimum? Not necessarily, because, as is evident from (6.7), the sensitivity of the measurement is proportional to the number of turns. Hence, we meet again the perennial problem of bandwidth versus sensitivity: their product usually is a constant for any given technique. Compromise (otherwise known as "trade off") is central to the system or device designer's art!

6.3.2 Direct Current Measurement

The fact that the Faraday magneto-optic effect allows the fiber measurement of direct current, in addition to alternating current, offers a potentially important advantage over conventional current-measurement transformers.

An investigation into the possibility of engineering a dc device has yielded some important additional features of the polarization behavior in monomode fibers.

The most notable of these was the apparently marked increase in the vibrationally induced polarization noise. Investigation showed that this was true only when the fiber was bent in more than one plane, and analysis soon revealed that this was a purely geometrical effect [5]. In essence, it is due to the fact that bending the fiber in one plane will rotate the direction of polarization of a propagating component linearly polarized in the plane of the bend, due to the necessity for the electric vector to remain normal to the fiber axis (Figure 6.14). If the fiber, additionally, is bent into another plane, the polarization direction of the light will depend upon the bend radius in the first plane (see Figure 6.14). Hence, the emergent polarization state will be dependent on the fiber geometry, and small vibrational effects can lead to relatively large variations in the output polarization direction.

As these geometrical effects will be optically reciprocal, the resulting noise may be removed by using the go-and-return reflection system described in the

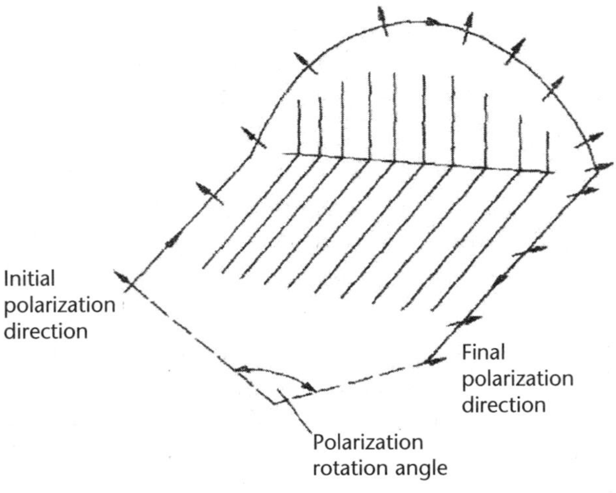

Figure 6.14 Geometrical rotation of polarization state.

last section. However, this may introduce *etaloning* effects that lead to drift in the dc device, and thus it now becomes necessary to use a low-coherence light source such as a multimode semiconductor laser.

Finally, it was also found that the polarization properties of such components as microscope objectives, polarizing prisms, optical filters, and so on, varied significantly with aperture position, so that small drifts in the launch geometry or source polar diagram (due, for example, to temperature-dependent stresses) caused corresponding drifts in the output polarization state of the light from the fiber. These drifts were large enough to be intolerable for a practical dc device. Thus, a successful dc device will require either a much better quality of discrete components, or a carefully engineered integrated optical launch package.

It follows, also, that alternating current measurement devices would benefit from a proper consideration of these new features in the engineering design, as the same disturbances will be present, even though they are less significant.

6.3.3 Voltage Measurement

Although the preferred method for optical-fiber measurement of current is the line integral of magnetic field using the magneto-optic effect, that for measurement of voltage is the line integral of electric field using the electro-optic effect [5].

The electro-optic effect (Pockels, Kerr) comprises an electric field-induced linear birefringence. The Pockels effect provides birefringence, which varies linearly with field, and can exist only in crystalline media, whereas the Kerr effect is a quadratic effect and can occur in both crystalline and noncrystalline media. In both cases, the effect is reciprocal, and thus contrasts with the magneto-optic effect. The quadratic nature of the Kerr effect precludes its use directly for line integration, but a convenient method for line integration using the Pockels effect in a fiber is shown in Figure 6.15.

A fiber which provides a linear birefringence proportional to the axial electric field is strung from Earth up to a high-voltage conductor.

Light which propagates along the fiber will thus experience two components of linear birefringence: an intrinsic component, due to the (necessary) crystal structure of the core, and the electric field-induced component, the latter allowing the required line integration of the electric field. Another similar fiber, with crystal axes rotated through 90°, is now joined to the first, and brings the light back down to Earth. Apart from the fact of bringing the light back down to a convenient Earthy measurement point, the reason for the second fiber is twofold: first, the intrinsic birefringence component of the fiber cancels that of the first, whereas the electric field-induced component adds to that of the first, as the electric field is now of the opposite sign; secondly, provided that the

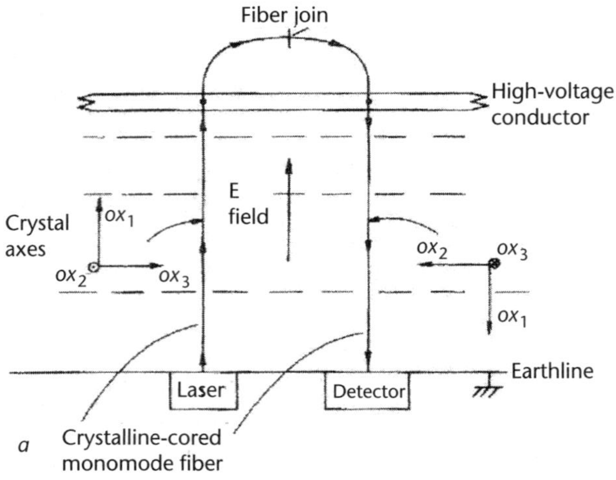

Figure 6.15 Optical-fiber voltage measurement.

fibers lie physically close together, vibrationally induced birefringence will also cancel in the two fibers. (Temperature effects should also cancel if the fibers lie close, as the intrinsic birefringence will then remain the same.)

Suitable fibers for the realization of this device do not yet exist: it is very difficult to fabricate crystalline-cored fiber. However, the method has been investigated experimentally using crystalline quartz rods, and has been shown to behave as expected [6].

A natural extension of these ideas is to combine electro-optic voltage measurement with magneto-optic current measurement (Figure 6.16). As the linear and circular birefringence components are fundamentally separable, both voltage and current indications are separately available, allowing, also, real- and reactive-power indications to be derived.

6.4 Distributed Polarimetric Sensing

Just as the one-dimensional nature of the fiber allows us to line-integrate within a measurand field, so it also allows us to line-differentiate. This then provides a measurement of the spatial distribution of the measurand field along the length of the fiber, so that we can observe its behavior in both space and time. To do this we must again time-resolve the continuously backscattered radiation from a propagating pulse, in order to obtain the spatial information.

6.4.1 Introduction to Distributed Optical-Fiber Measurement

Optical-fiber distributed measurement sensing is another technique which utilizes the one-dimensional nature of the optical fiber as a distinct measurement

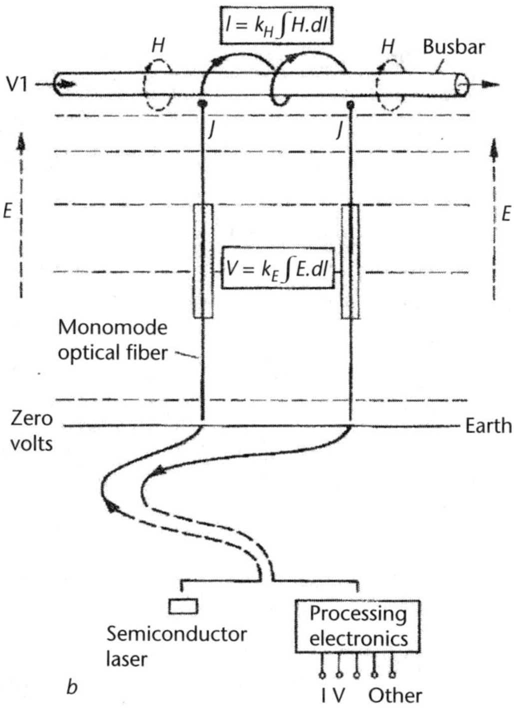

Figure 6.16 Combined current and voltage measurement.

feature. It is possible, in principle, to determine the value of a wanted measurand continuously as a function of position along a length of a suitably configured optical fiber, with arbitrarily large spatial resolution. The normal temporal variation of the distribution is determined simultaneously.

Such sensing systems are normally referred to as fully distributed systems, to distinguish them from the quasi-distributed systems, which possess the capability of sensing the measurand only at a number of discrete, predetermined points.

The fully distributed facility opens up an enormous number of possibilities for industrial application. For example, it would allow the spatial and temporal strain distributions in large critical structures, such as multistory buildings, bridges, dams, aircraft, pressure vessels, electrical generators, and so on, to be monitored continuously. It would allow the temperature distributions in boilers, power transformers, power cables, aerofoils, office blocks, and so on, to be determined, and thus heat flows to be computed. Electric and magnetic field distributions could be mapped in space so that electromagnetic design problems would be eased and sources of electromagnetic interference would be quickly identifiable.

There are two important definable reasons for requiring the information afforded by distributed optical-fiber measurement sensors. The first is that of providing continuous monitoring so as to obtain advance warning of any potentially damaging condition in a structure, and thus to allow alleviative action to be taken in good time. The second is that this spatial and temporal information allows a much deeper understanding of the behavior of large (or even quite small) structures, with many implications for improvements in their basic design.

Conventional industrial measurement sensor technology doesn't provide this facility. When measurand distributions of any kind are vital in a given situation, the solution usually is to festoon the structure with a multitude of thermocouples, or strain gauges, or whatever. This then presents problems of multiplexing, logging, and calibration, and, in any case, relies on the choice of position for each of the many sensors being the correct one—a choice that cannot properly be made without a prior knowledge of the very distribution one is seeking to measure! This "solution" is thus expensive, tedious, and usually broadly inadequate.

The optical fiber, on the other hand, can be readily installed in industrial plant (retrospectively if necessary), produces minimal disturbance of the measurement environment, is cheap, passive, and electrically insulating, acts as its own telemetering channel, can easily be rearranged in accordance with acquired knowledge, and allows a choice of any or all measurement points along its length within the limits of the spatial resolution interval. If such a technique can be made to work satisfactorily for a number of measurands, a new dimension appears in the field of industrial measurement. These possibilities are being actively explored.

A good polarimetric example of this technique will now be described [6].

6.4.2 Polarization-Optical Time Domain Reflectometry (POTDR)

In Section 5.4.5 we learned how it was possible to measure the spatial distribution of PMD along the length of a monomode fiber using the light Rayleigh-backscattered from a propagating pulse. The relations given by (5.9a), (5.9b), and (5.9c) express the spatial variations of the defining polarization parameters of the fiber, $\delta(z)$, $q(z)$, $\rho(z)$, as a function of the parameters $\delta_e(z)$, $q_e(z)$, and $\rho_e(z)$ defining the polarization properties of the retarder/rotator pair equivalent to the length of fiber up to distance z.

These equations are restated here, for easy reference:

$$\delta(z) = ((\partial\delta_e/\partial z)^2 + \sin^2 2\delta_e(\partial q_e/\partial z)^2)^{1/2}$$
$$q(z) = q_e + \rho_e + 1/2 \cdot \tan^{-1}(\sin 2\delta_e(\partial q_e/\partial z)/(\partial\delta_e/\partial z))$$
$$\rho(z) = \partial\rho_e/\partial z + 2\sin^2\delta_e(\partial q_e/\partial z)$$

Clearly, if $\delta(z)$, $q(z)$, and $\rho(z)$ are known, then so is the distribution of any external field which modifies these parameters, such as strain, temperature, displacement, electric field, magnetic field, and so on. However, there remains the problem implicit in these latter two equations, namely that ρ_e is unmeasurable in backscattered radiation: it is cancelled in the go-and-return process (provided that the circular birefringence is reciprocal; i.e., is not due to a magnetic field). However, the first of these equations does not contain ρ_e, so its distribution can be determined directly from a knowledge of the distributions of the measurables δ_e and q_e.

The first step in tackling the more general problem of determining all of $\delta(z)$, $q(z)$, and $\rho(z)$ is to differentiate the second equation with respect to z and then to subtract it from the last equation so as to eliminate $\partial\rho_e/\partial z$. This gives

$$\rho(z) - \partial q(z)/\partial z = \Phi(\delta_e(z), q_e(z)) \tag{6.11}$$

where Φ, clearly, is a defined function only of the measurables $\delta_e(z)$, $q_e(z)$.

This is a useful result when, as is usually the case, the circular birefringence is a consequence only of fiber twist, because $\partial q/\partial z$, the rotation of the linear birefringence axes with z, represents exactly this twist. Indeed, we may write, to a very good approximation:

$$\rho(z) = k\partial q(z)/\partial z$$

where k is a known constant (Section 4.2.4). So that (6.10) becomes

$$(k - 1)\partial q(z)/\partial z = \Phi(\delta_e(z), q_e(z))$$

and hence both $\partial q(z)$ and $\rho(z)$ are known.

If we cannot assume that the circular birefringence is due only to fiber twist, then more drastic action is necessary.

For this, as was the case for the measurement of PMD distribution, we use the fact that $q(z)$, being a physical direction in space, is independent of optical wavelength, λ.

Differentiating, now, (6.11) with respect to λ we have

$$\partial\rho(z)/\partial\lambda = \partial\Phi(\delta_e(z), q_e(z))/\partial\lambda \tag{6.12}$$

We now use the known physical form of the dependence of the circular birefringence on wavelength. To first order we have, for any uniform element l,

$$\rho = (2\pi/\lambda) \cdot \Delta n_c l \tag{6.13}$$

where Δn_c is the difference in refractive index between right-hand and left-hand circularly polarized eigenmodes (Δn_c will be weakly wavelength dependent, which is why the equation is true only to first order. However, the wavelength dependence is known, so that calculation could be performed to higher orders if necessary). Hence, from (6.12) we have

$$\partial\rho(z)/\partial\lambda = -(2\pi/\lambda^2) \cdot \Delta n_c l = -\rho(z)/\lambda \tag{6.14}$$

Thus, from (6.12) and (6.14) we have

$$\rho(z) = -\lambda\partial\Phi(\delta_e(z), q_e(z))/\partial\lambda$$

and $\rho(z)$ becomes known, together with $\partial\rho_e(z)$ from the last of the Chapter 5 equations listed above. $\partial q(z)/\partial z$, also is known, so that $q(z)$ can be recovered by integration, if necessary.

Hence we can, in principle, recover all of $\delta(z)$, $q(z)$, and $\rho(z)$ provided that the distributions of the Stokes parameters are known as a function of optical wavelength.

Clearly, however, there are severe problems in regard to the accuracy of the determinations in the face of system noise.

Figure 6.17 shows the experimental arrangement for an implementation of POTDR which measures only $\delta(z)$, and hence can be performed using only one wavelength. A pulse of polarized light is launched into a fiber and the Rayleigh back-scattered radiation is polarization-analyzed by a three-way splitting and subsequent simultaneous measurement of the Stokes parameters, in real time. Hence, the instantaneous polarization state of the returning light is known for the complete passage of the pulse along the fiber. This allows $\delta(z)$ to be determined when the 40m of fiber at the far end of a 2-km fiber length are wound on an 80-mm diameter coil; the result of the determination is shown in Figure 6.18. The coil winding endows a bend birefringence whose value and position are measured.

Another useful application is shown in Figure 6.19. Here a section of side-hole fiber [Figure 6.19(a)] is subjected to a gas pressure in a pressurized cylinder. The isotropic pressure acts upon the asymmetrical fiber structure to endow upon the core a linear birefringence proportional to the pressure. The result is shown in Figure 6.19(b), where it is seen that the pressure must first act to cancel the intrinsic linear birefringence of the fiber. The linearity of the result indicates the value of this technique in measuring the distribution of any fluid pressure, such as that in oil wells or water pipes [7].

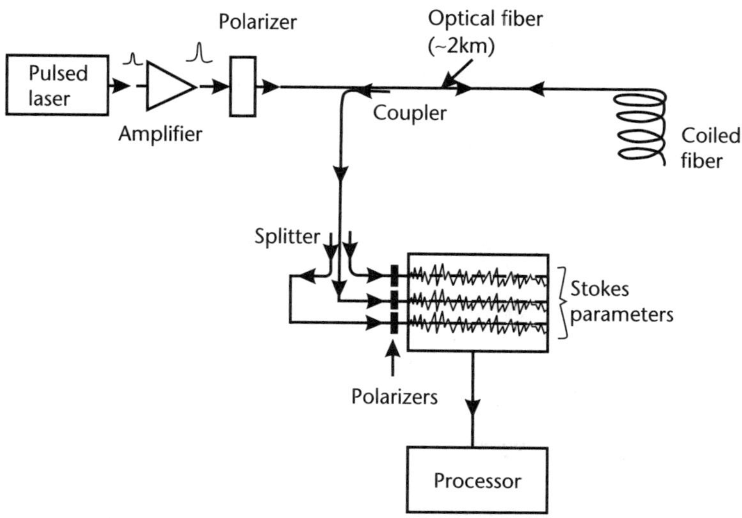

Figure 6.17 POTDR determination of the distribution of linear birefringence, $\delta(z)$.

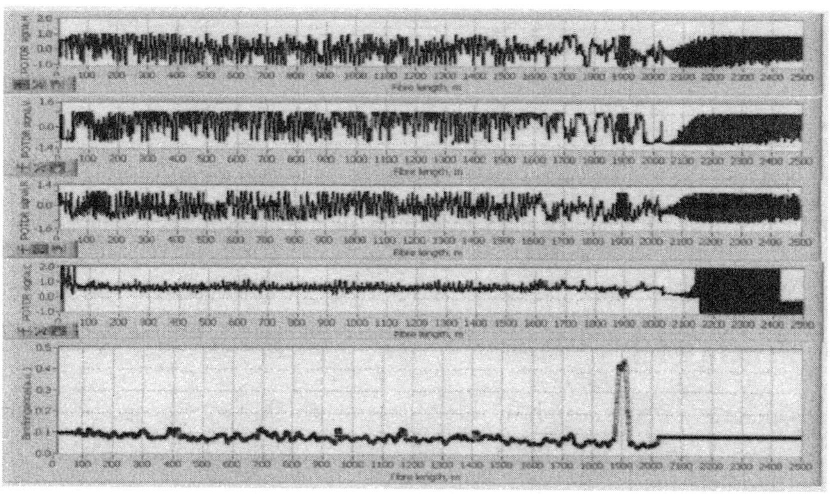

Figure 6.18 Stokes parameter traces and processed output for the linear birefringence distribution: far-end coiled fiber. (*From:* [6]. © 2002 IEEE. Reprinted with permission.)

The more general use of CPOTDR can be extended to any structure whose strain field needs to be monitored, such as dams, bridges, roads, aircraft, spacecraft, power transformers, or industrial pressure vessels. The technique is unique in its measurement of transverse ("squeeze") strain (linear birefringence) and shear strain (circular birefringence/twist) [8].

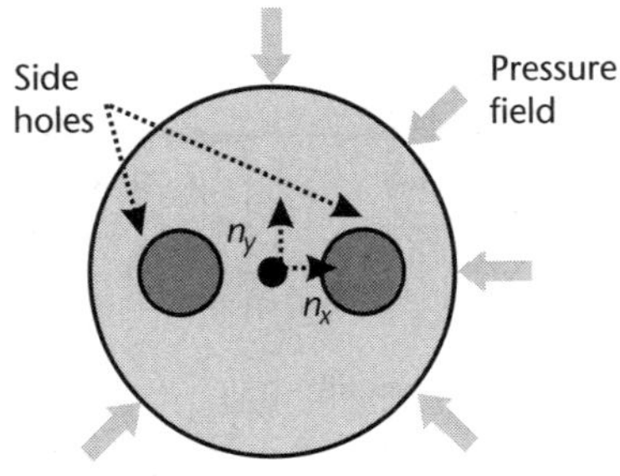

(a)

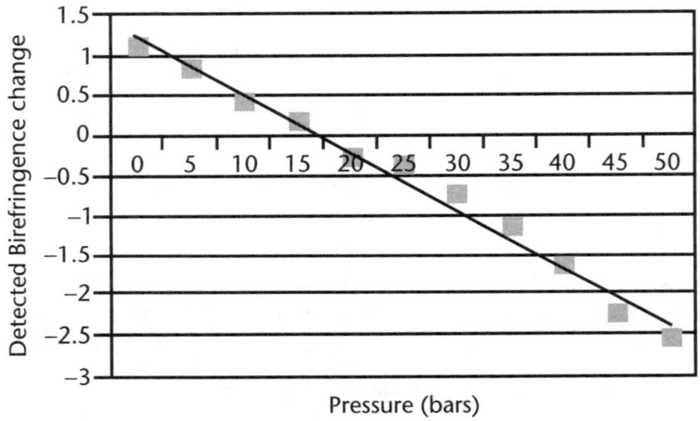

(b)

Figure 6.19 POTDR gas-pressure measurement: (a) side-hole fiber; and (b) measured birefringence/pressure relationship. (*From:* [9]. © 2005 IEEE. Reprinted with permission.)

6.5 Conclusions

In this chapter we have seen how the "directionality" of the polarization properties of optical fibers can be useful in the measurement-sensing function. To the extent that external fields can alter deterministically these transverse directional properties, they can be measured. The fiber is especially useful in this measurement function for three reasons:

1. The optical interaction paths with the measurand field can be long, leading to good (and tailorable) sensitivity.
2. The fields can be integrated along the length of the fiber when appropriate (e.g., for measurement of current or voltage).
3. The field can be differentiated along the length of the fiber, leading to distributed measurement (e.g., strain and temperature profiles on large, critical structures).

References

[1] Dakin, J. P., and B. Culshaw, (eds.), *Optical Fiber Sensors: Volumes 1–4*, Norwood, MA: Artech House, 1988–1997.

[2] Grattan, K. T. V., and B. T. Meggitt, (eds.), *Optical Fiber Sensor Technology*, London, U.K.: Chapman and Hall, 1995.

[3] Sudo, S., *Optical Fiber Amplifiers*, Norwood, MA: Artech House, 1997.

[4] Lefevre, H., *The Fiber-Optic Gyroscope*, Norwood, MA: Artech House, 1993.

[5] Rogers, A. J., "Polarization Optics for Monomode Fiber Sensors," *Proc. IEE*, Vol. 132, Pt. J, No. 5, October 1985, pp. 303–308.

[6] Rogers, A. J., "Distributed Fiber Measurement Using Backscatter Polarimetry," *Proc. Optical Fiber Sensors (OFS) 2002*, Portland, OR, May 2002, pp. 367–370.

[7] Kanellopoulos, S. E., S. V. Shotolin, and A. J. Rogers, "Method and Apparatus for Detecting Pressure Distribution in Fluids," International Patent Application PCT/GB20005/002050, 2005.

[8] Rogers, A. J., "Distributed Measurement of Strain Using Optical-Fiber Backscatter Polarimetry," *Strain*, Vol. 36, No. 3, August 2000, pp. 135–142.

[9] Rogers, A. J., S. V. Shotolin, and S. E. Kanellopoulos, "Distributed Measurement of Fluid Pressure Via Optical-Fibre Backscatter Polarimetry," *Proc. OFS 17*, Bruges, 2005, pp. 230–233.

7

Applications of Nonlinear Polarization Effects in Optical Fibers

7.1 Introduction

Various nonlinear polarization effects that occur in optical fibers were described in Section 4.3. We noted there that fibers were quite efficient generators of nonlinear optical effects, owing to their ability to maintain a large optical intensity over long distances. In this short chapter we shall examine some of the practical consequences of nonlinear polarization effects for the kinds of optical measurement sensing and telecommunications systems we have considered in previous chapters.

The consequences of the nonlinear effects result largely from interactions either within an optically guided wave, or between separate waves. The electric field of a wave can alter the refractive index of the fiber core (as a result of the atomic distortions to which it gives rise), via one or more of the nonlinear effects, thus altering the propagation conditions either for itself, or for another wave propagating through the same region simultaneously.

The ideas will become clearer as we consider some specific examples. We begin with a point sensor.

7.2 The Optical Kerr Effect in the Optical-Fiber Gyroscope

It was noted, when considering the action of the optical-fiber gyroscope in Section 6.2.1, that we require the paths of the two counterpropagating wave components, ideally, to be identical under the condition of zero rotation. This will ensure that the resulting interference pattern will be dependent only on the rotation itself.

However, the electric field of one beam will act, via the optical Kerr effect, to alter the phase of the other. The effect is small but then so is the phase difference which we are seeking to measure, at low rotation rates. We can calculate this effect, using the ideas discussed in Section 4.3.6. We know that in fused silica the nonlinear electric polarization can be written:

$$P(E) = \chi_1 E + \chi_3 E^3 \tag{7.1}$$

to a good approximation.

The electric field, in this case, will be given by the sum of two counter-propagating waves:

$$E = E_+ \exp i(\omega t + kz) + E_- \exp i(\omega t - kz)$$

Substituting this value into (7.1) gives

$$P(E) = \chi_1 E + P_{N+} + P_{N-}$$

where

$$P_{N+} = \chi_3\left(E_+^2 + 2E_-^2\right)E_+ \exp i(\omega t + kz)$$

$$P_{N-} = \chi_3\left(E_-^2 + 2E_+^2\right)E_- \exp i(\omega t - kz)$$

Now the refractive index is given in general by

$$n^2 = \epsilon = 1 + \chi = 1 + \frac{P(E)}{E}$$

Hence for the clockwise (+) propagating beam we have

$$n_+ = (1 + \chi)^{1/2} = \left(1 + \chi_1 + \frac{P_{N+}}{E_+ \exp i(\omega t + kz)}\right)^{1/2}$$

or, using the binomial theorem,

$$n_+ = n_0 + \frac{1}{2}\chi_3\left(E_+^2 + 2E_-^2\right)$$

Similarly,

$$n_- = n_0 + \frac{1}{2}\chi_3\left(E_-^2 + 2E_+^2\right)$$

It follows that the nonlinear (optical Kerr effect induced) phase changes for each direction are

$$\Delta\varphi_+ = \frac{2\pi}{\lambda_0} l \frac{\chi_3}{2}\left(E_+^2 + 2E_-^2\right) \tag{7.2}$$

$$\Delta\varphi_- = \frac{2\pi}{\lambda_0} l \frac{\chi_3}{2}\left(E_-^2 + 2E_+^2\right)$$

Clearly, these are not the same unless $E_+^2 = E_-^2$: the difference is, in fact,

$$\Delta\varphi_+ - \Delta\varphi_- = \left(E_-^2 - E_+^2\right)\frac{\pi}{\lambda_0} l\chi_3$$

A difference in optical power of just 1 μW leads to a phase discrepancy of 10^{-6} μrad, equivalent to a rotation of 0.01° h^{-1}, whereas these devices are actually required to measure ~ 0.01° h^{-1}. This difference in optical power can easily result from the fiber attenuation, which produces inequalities of power, away from the center of the coil.

The problem can be overcome by square-wave modulating the laser power. In this case each beam is influenced by the other for only half the time, so that, effectively, the cross-product term in (7.2) is reduced by a factor of 2 in each case, now giving:

$$\Delta\varphi_+ = \Delta\varphi_-$$

which is, of course, the required condition.

Other sources of noise are: Rayleigh backscatter in the fiber, which gives a coherent interfering signal, and drift in the value of the area of the fiber, due to temperature variation.

7.3 Nonlinear Distributed Sensing Methods

7.3.1 General

Polarization techniques are generally very powerful in pursuit of optical-fiber sensing that is both sensitive and practical; and this applies to both point and distributed sensors. In order to utilize, for measurement purposes, the effect that measurands produce on the polarization properties of optical fibers, it is highly desirable for the intrinsic polarization properties of the fiber to be well defined in advance. To this end (and for others) we remember that fibers with particular polarization properties have been developed: notable amongst these are those with large values of linear birefringence (hi-bi) and of circular birefringence (hi-ci-bi) (see Section 4.2.2).

We shall describe three nonlinear polarization-optical methods for implementing forward-scatter (as opposed to backscatter) DOFS. The first two are *dynamic* methods that utilize the optical Kerr effect in the fiber; the third employs a *static* technique where photo-refractive gratings are written into the fiber using the well-known phenomenon of photo-sensitivity in certain optical materials [1].

Another distinction to be noted is that the dynamic methods are fully distributed, allowing measurement to be made at any point along the fiber, while the static method is quasi-distributed, with measurement possible only at certain, prescribed points along it.

7.3.2 Frequency-Derived Distributed Optical-Fiber Sensing (FD/DOFS)

The first dynamic method makes use of a coupling grating, which propagates at the speed of light in a hi-bi fiber. Suppose that a narrow, linearly polarized pulse of light is launched with its polarization direction at 45° to the fiber's eigenaxes (Figure 7.1). This launches equal light components into each of the two eigenmodes. Since velocities differ for the two modes, the resultant polarization state varies cyclically from linear to circular to orthogonal-linear to circular and back to the original state (Figure 7.1). The distance over which

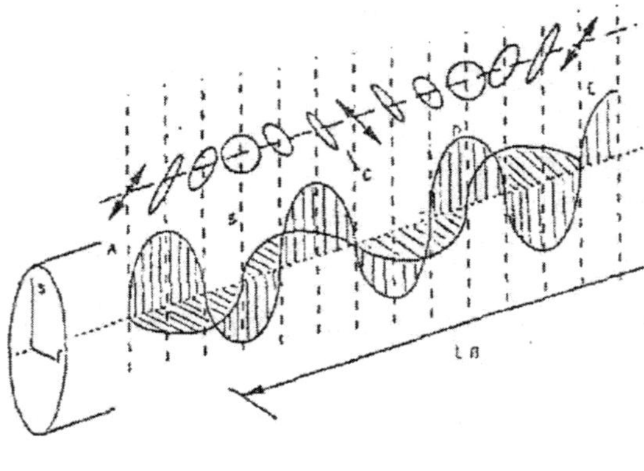

Figure 7.1 Evolution of polarization state along a hi-bi fiber. (*From:* [2]. © 1993 SPIE. Reprinted with permission.)

a complete cycle occurs within the fiber is known as the beat length, and is constant for a given wavelength: it is typically a length of a few millimeters.

The relationship between birefringence and wavelength is important, and easily derived. The birefringence (B_I) is simply the difference in the refractive indices (Δn) for the two polarization directions, and the beat length, L_B, is given by

$$2\pi/\lambda \cdot \Delta n L_B = 2\pi, \text{ that is, } L_B = \lambda/\Delta n = \lambda/B_I \qquad (7.3)$$

where λ is the wavelength of the light.

The optical Kerr effect (see Section 4.3.6) is that effect whereby the refractive index of a material is altered by the action of the electric field of a propagating wave. The refractive index is increased for light polarized in the direction of the electric field compared with the orthogonal linear polarization. The magnitude of the change is given by

$$B_k = \lambda_0 b E^2 \qquad (7.4)$$

where λ_0 is the free-space wavelength, b is the Kerr coefficient, and E is the optical electric field.

Consider now the optical Kerr effect of the narrow pulse propagating in the hi-bi fiber. Its effect will be to induce an extra birefringence in the fiber, and this will be a maximum when the pulse light is linearly polarized, occurring at intervals of one-half beat length. The induced axes at the maxima will be at $\pm 45°$ to the intrinsic axes and will cause the resultant birefringence axes to rock from side to side, at the half beat length interval, and through a small angle which is approximately equal to the ratio of induced to intrinsic birefringence values.

Consider now a cw light wave with the same wavelength as the pulse light, counter-propagating against the pulse, and linearly polarized along one of the birefringence axes. When it encounters the pulse, it will be coupled, by the rocking of the axes, to the orthogonal eigenmode [2]. Clearly the coupling will be at a maximum when the pulse is linearly polarized, so that the coupled light is amplitude modulated at a frequency,

$$f_D' = c/L_B \qquad (7.5)$$

where c is the velocity of light in the fiber and L_B is the fiber beat length. Thus, the coupled cw light (in the eigenmode orthogonal to that of its entry into the fiber) will emerge from the fiber continuously amplitude-modulated at this frequency. The "instantaneous" frequency of the amplitude modulation at any given time will correspond to the value of the beat length over the region in the fiber corresponding to the position of the pulse at that time. Thus, a measurement of f'_D as a function of time will map the fiber's birefringence as a function of position and, consequently, will correspondingly map the distribution of any external field, such as temperature or strain, whose action modifies the local value of birefringence. Thus we have a distributed optical-fiber sensor operating in forward-scatter, and enjoying the advantages of a much larger signal level (compared with a backscattered signal) and a frequency measurement independent of the light power level and its practical vagaries.

However, one problem in regard to the practical implementation of FD/DOFS is immediately apparent: for an intrinsic fiber beat length of, typically, 2 mm, (7.5) gives $f'_D = 100$ GHz. This is too large for convenience in a practical detection and processing system: no practical photo-detector is yet that fast. (A small beat length is required in order to achieve satisfactory polarization holding; that is, polarization coupling should be due almost exclusively to the optical Kerr effect, and not to capricious external agencies.) This problem may be resolved very conveniently by operating the pulse, henceforward referred to as the optical "pump," and the cw, referred to as the optical probe, at different wavelengths, λ_1 and λ_2, say, respectively.

Suppose that the pump pulse is launched, as before, with its linear polarization direction at 45° to the axes, and that the probe is also launched, from the other end of the fiber, at 45° to the axes. This time the probe light also undergoes a cyclic variation in polarization state but now with a different beat length from the pump, since its wavelength is different (7.3). Maximum coupling of the probe now occurs when both pump and probe are linear at the same points in the fiber (NB the axis-rocking angle is still very small compared with 45°). Such points are separated by a distance L_e (effective beat length) where

$$\frac{1}{L_e} = \frac{1}{L_1} - \frac{1}{L_2}; \; L_e = \frac{\lambda_1 \lambda_2}{\Delta\lambda B_I} \; (\Delta\lambda = \lambda_1 - \lambda_2)$$

L_1 and L_2 being the beat lengths for the pump and probe waves, respectively.

Thus, there is added to the fundamental coupling component at the high frequency (c/L_B) another component at much lower frequency, given by

$$f_D = c/L_e = cB_I \cdot \Delta\lambda/\lambda_1\lambda_2 \tag{7.6}$$

It is clear from (7.6) that measurement of f_D still allows B_I to be measured unambiguously and thus also the distribution of any parameter that modifies it.

To fix ideas as to the value of this frequency, we again take a 2-mm beat length at a wavelength of 1,000 nm, assume that $\lambda_1 \approx \lambda_2$, and find that

$$f_D \approx 10^{17} \cdot \Delta\lambda$$

Hence for $\Delta\lambda = 1$ nm, $f_D' \sim 100$ MHz. This is, clearly, now a much more reasonable value from the point of view of practical frequency-derived distributed measurement.

However, we must also concern ourselves with the strength of the coupling, for we obviously need an easily detectable signal (i.e., well above noise) if we are to measure accurately the frequency of its amplitude modulation. The coupling strength will depend upon pump energy, because it will be a function of the strength of the electric field and the length of time for which it acts on the cw wave.

The setup for FD/DOFS is shown in Figure 7.2(a), and a typical output trace is shown in Figure 7.2(b). The experimentally observed and theoretically calculated relationships between the frequency and wavelength shift are shown in Figure 7.2(c). This figure shows the expected linear trend connecting the derived frequency with the wavelength shift. Figure 7.3 shows the theoretical and experimentally observed relationships between the coupling strength and the pump pulse energy for two different types of hi-bi fiber.

Using the novel technique presented here, the spatial variation of the birefringence of a polarization-maintaining fiber can be measured remotely in a short time, and, since the signal is in the form of a frequency, it is immune from the common error sources present in intensity-coded systems.

7.3.3 Polarization State Dependent Kerr Effect Forward-Scatter DOFS

In this second Kerr effect forward-scatter method, the emphasis is on spatial location of an external perturbation rather than on the measurement of its magnitude, although suitable processing is capable, in principle, of revealing the latter. It is capable of providing good spatial resolution and rapid response for application to, for example, intrusion monitoring or vehicle location.

Figure 7.4(a) is a reminder for the action of the "normal" electro-optic Kerr effect. The optical Kerr effect arrangement for our present purposes, shown schematically in Figure 7.4(b), employs a length of polarization-maintaining fiber carrying two counter-propagating beams. A cw probe beam is launched from one end of the fiber so as to excite equally the two eigenmodes, and the polarization state of this beam is detected at the far end of the fiber by means of a beamsplitter and analyzer oriented at 45° to the birefringence axes. An intense, pulsed, pump beam is launched on one of the birefringence axes. The pump pulse causes a phase shift between the eigenmodes of the probe beam, leading to a change in the output polarization state of the probe. This is detected

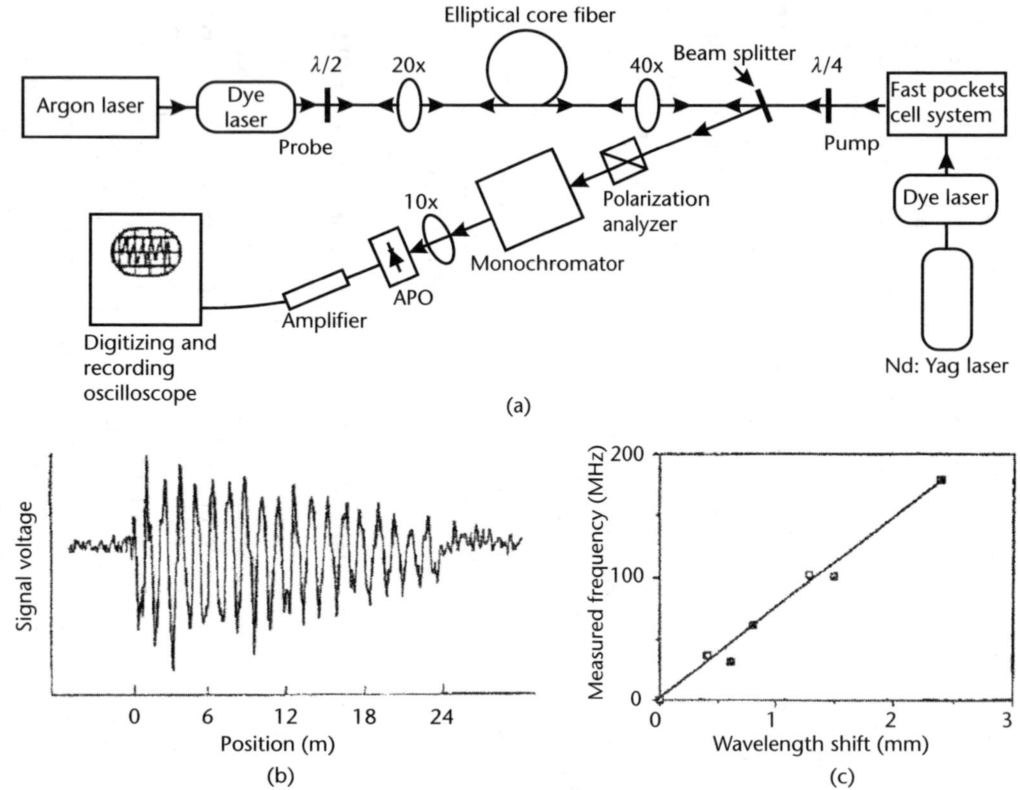

Figure 7.2 (a–c) Frequency-derived, forward-scatter DOFS. (*From:* [2]. © 1993 SPIE. Reprinted with permission.)

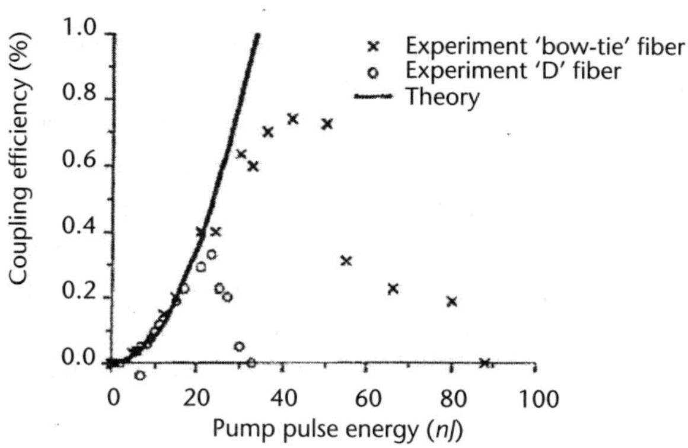

Figure 7.3 Coupling efficiency versus pump energy for FD/DOFS. (*From:* [2]. © 1993 SPIE. Reprinted with permission.)

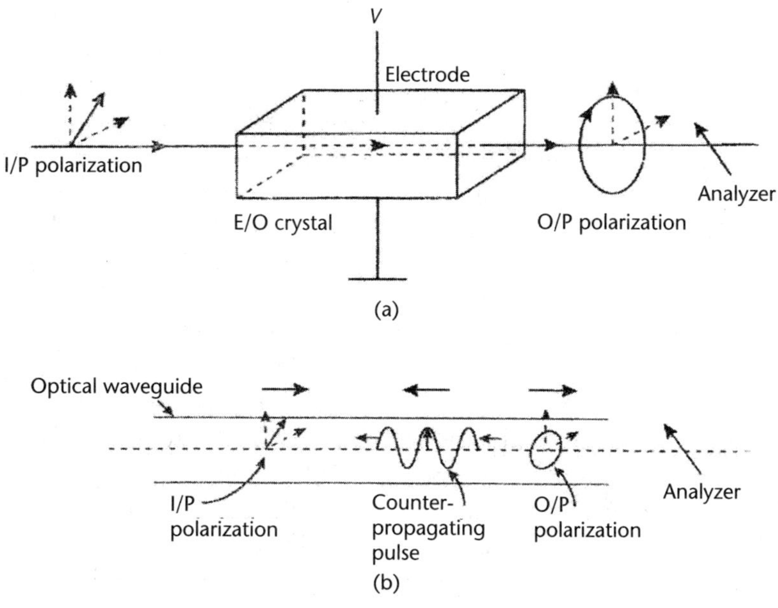

Figure 7.4 (a) Normal electro-optical Kerr effect. (b) Optical Kerr effect: light acting on light.

as a sharp change in the probe intensity passed by the analyzer when the pump is initially launched into the fiber.

If, now, a force acts at an angle to the axes along a section of fiber, coupling of the pump light to the other axis will occur, and the Kerr effect on the probe will thus be modified. The probe light itself will also experience mode coupling, which will further modify the output polarization state. The actual change which occurs will depend, inter alia, on the states of polarization of the beams as they enter the perturbed region and thus, unless the birefringence perturbation is very small compared with the intrinsic birefringence, there will exist a mutual dependence of effects from different measurement locations, which only fairly complex signals processing would be able to resolve. However, it is clear that for any change in the direction of birefringence axes consequent upon the perturbation by a measurand, there will, in general, result a change in optical Kerr effect. A differentiated signal, thus, will at least indicate differential features of the measurand distribution, even though a fully quantified spatial distribution is more elusive.

The fiber used in one experiment [3] was monomode high-birefringence fiber, with a diameter of 67 μm, attenuation of 35 dB/km at 633 nm, and core to cladding refractive index difference $\Delta n = 0.032$. The length of the fiber was about 100m. The schematic for the experiment is just that shown in Figure 7.4(b). Pump pulses (617 nm) of 8-ns (FWHM) duration were generated in a dye laser with a repetition rate of 50 Hz. These pump pulses were launched on to one of the birefringence axes of the fiber with the help of a half-wave plate and with a peak power of 3W measured at the output end of the fiber. The linearly polarized probe beam of wavelength 633 nm, from a He-Ne laser, was launched into the fiber at 45° to the birefringence axes. On emergence, the probe beam was directed by a beamsplitter to the detector, via the polarization analyzer; its average power at the detector was about 25 μW. The He-Ne laser and the detector were protected from the pump light by use of band-pass filters at 633 nm. Force was exerted by pressing metal rods on the fiber. The received signals were recorded, averaged, and differentiated using the functions of the digital storage oscilloscope.

In the absence of any measurand-induced perturbation, the Kerr effect of the pulse is to modify the local value of birefringence as it propagates, a modification which is sensed by the probe beam as a phase shift between the eigenaxes, and which is, in principle, constant for the duration of the pulse's passage through the fiber. The effect of this phase shift on the optical signal passing through the analyzer is shown in Figure 7.5(a). In practice, the slow fall in the value of the phase shift is due to the attenuation of the pump pulse with distance along the fiber. Figure 7.5(b) also shows the effect of differentiating this signal with respect to distance.

Figure 7.5(c) shows the fluctuating analyzer signal when the fiber was perturbed at two points, and its differential with respect to distance. The points at which the weights were applied are clearly evident. Such a system, even as it stands, could be used as an intruder alarm or as an indication of anomalous disturbance of almost any kind.

Of course, it is true that if the perturbing force acts along one of the eigenaxes, no rotation of the axes occurs and there is no resultant polarization

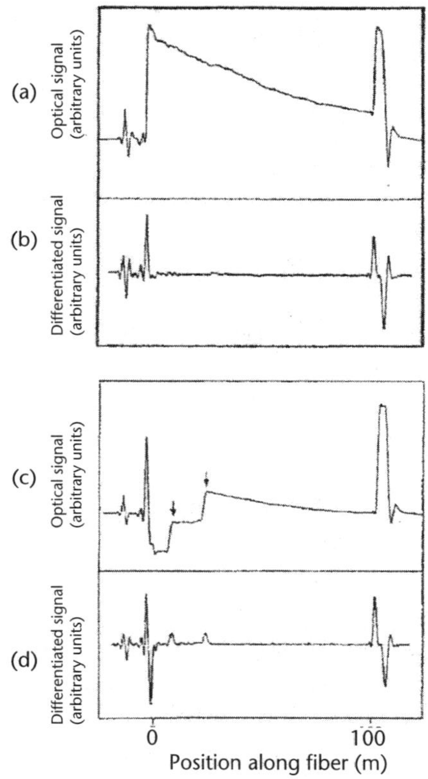

Figure 7.5 (a–d) Traces for Kerr effect DOFS. (*From:* [3]. © 1992 SPIE. Reprinted with permission.)

perturbation. Hence, either the direction of the perturbing force must be known (e.g., a vertical weight) or two fibers may be used: these should run closely parallel and be orientated with their birefringence axes at 45° to each other.

Thus is demonstrated the use of the optical Kerr effect to determine the locations of discrete mode coupling points spaced along a polarization-maintaining fiber. Differentiation of the received signal with respect to time provides a simple way to reduce confusing interactions when multiple coupling points are present.

7.3.4 Quasi-Distributed Sensing Using Photo-Induced Polarization Grating Couplers

7.3.4.1 A Review of Fiber Bragg Gratings (FBGs)

Photosensitivity was discovered in optical fibers in 1978 [1]. This is a phenomenon whereby the refractive index of a fiber material (i.e., doped silica) can be modified (permanently or semi-permanently) by exposure to ultraviolet (UV) light. The mechanisms of the effect are various, complex, and as yet incompletely understood, although it is known that they involve either electron traps created by the impurities in the material structure (type I, nonpermanent), or actual physical damage to the core–cladding interface (type II, permanent). However, the methods by which photosensitivity can be induced in fibers are now well tried and tested.

Photosensitivity effects in optical fibers are well documented. The holographic method [4] for writing controlled Bragg gratings in germanium-doped fibers has led to the possibility of quasi-distributed measurement systems capable of examining either temperature or stress/strain distributions. The writing is done by setting up an interference pattern along the fiber core, using UV light (Figure 7.6). The light acts photo-chemically to modulate the refractive index, longitudinally, according to the impressed pattern. Hence, a linear sinusoidal variation of the index is written along the fiber, over a distance that can be varied from a few millimeters to tens of centimeters. Optical spectrum analyzers (OSAs) monitor the development of the grating, and an index-matching cell (IMC) prevents back-reflections from the coupler. In this way a series of gratings with different grating spacings can be written along the length of a fiber, and the individual gratings are identified in the optical frequency domain using either a broadband optical source or a swept-optical-frequency source. Various interrogation schemes have been suggested for this type of arrangement.

Optical-fiber gratings are finding a variety of applications as components in optical communications systems (e.g., wavelength filters, selective reflectors in fiber lasers, and dispersion compensators) but they are also extremely useful in quasi-distributed optical-fiber sensors (QD/DOFS) systems. An illustration of their use is shown in Figure 7.7. Here, a number of sinusoidal fiber Bragg gratings is arranged along a monomode fiber, each grating selectively reflecting

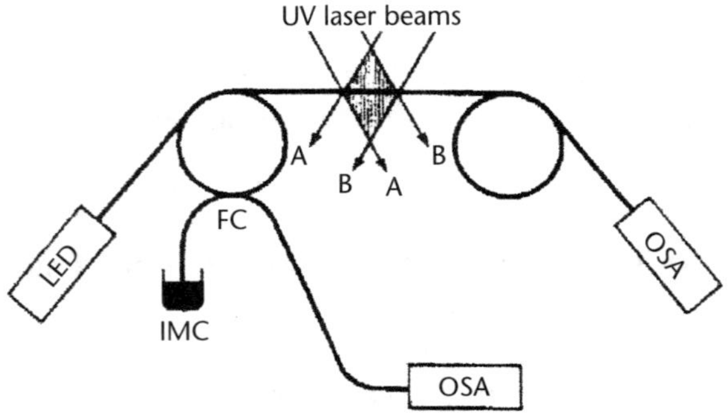

Figure 7.6 Writing a fiber Bragg grating. (*From:* [4]. © 1984 OSA. Reprinted with permission.)

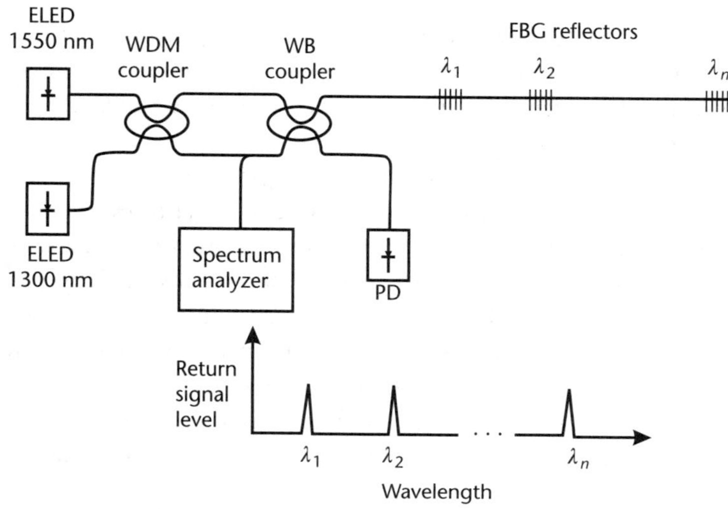

Figure 7.7 Linear array of FBGs for quasi-distributed measurement. (*From:* [2]. © 1993 SPIE. Reprinted with permission.)

a different wavelength according to its spatial period. The center wavelength of the reflection is given by

$$\lambda_B = 2n\Lambda$$

where n is the refractive index of the fiber material and Λ is the spatial period (the "grating spacing"). If the system is now interrogated with a broadband source (edge-emitting, light-emitting diode: ELED), the reflection spectrum consists of a series of peaks, each one corresponding to a particular grating, which is thus thereby identified. Now λ_B is dependent upon external fields that may vary n or Λ. The most important of these are temperature and strain, each of which modifies both n and Λ, and which thus can be measured quasi-distributedly. The writing process for the gratings gives easy control over their length, which can be as small as 10 to 20 mm. Hence, the spatial resolution is now of this order, which is very high by DOFS standards. Clearly, as the number of gratings along the fiber increases, the design of an effective interrogation system becomes more difficult. The source cannot be too broad in bandwidth, for this creates difficulties with regard to the launching of power into the fiber and to the fiber's attenuation spectrum and dispersion characteristics. If a single source is scanned in wavelength, the scanning range will be limited. There will also be difficulties with regard to the wavelength analysis of the returning light. Considerable attention has been paid to this problem, and the present limitation is at about 30 gratings. The durability, flexibility, and versatility of fiber gratings will ensure that development in this area will continue and that this technology will have a significant role to play in the monitoring and diagnostics of strain and temperature in extended structures well into the future. Already there have been many successful field trials.

QD/DOFS systems are valuable in that, by presensitizing specific regions of the fiber to the wanted measurand field, several advantages are conferred: first, the sensitization can be highly specific to the wanted measurand, leading to a good SNR at the detector and thus a good measurement sensitivity; secondly, the sensitive region can be made arbitrarily small, leading to good spatial resolution; thirdly, the sensitized regions (transducers) can be encoded to allow easy identification at the detector (the particular grating period of a fiber grating leading to a known frequency of optical reflection is a good case in point) and thus to simple detection-electronics. However, there are disadvantages: since the transducer points are fixed, the crucial regions in the measurand field must be known in advance, and must remain fixed. This is by no means always convenient. Furthermore, the sensitization of the fiber at fixed points usually leads to large attenuation of the interrogating light and thus to limited numbers of transducers, and limited dynamic range. Finally, the necessity for decoding, in addition to other, more standard, requirements, tends to lead to a burdensome complexity in the electronics.

Clearly, a series of Bragg gratings such as this can act as a quasi-distributed measurement system for temperature alone in a strain-free arrangement, or simultaneously for temperature and strain, provided that the two can be discriminated, and much attention has been given to effective discriminatory procedures. These include the use of gratings with different spacings overlaid at the same position, the use of two spatial propagation modes, and the use of two polarization modes. It is to this last method that we now turn our attention.

7.3.4.2 Discrimination Using Polarization Rocking Filters

A primary difficulty with FBGs is that they are sensitive simultaneously both to temperature and to strain, so that if one of the parameters is to be measured the other must be known.

A method for overcoming this problem uses hi-bi fiber. In this method, use is made of a photo-induced polarization rocking filter. Such filters are produced by exposing a hi-bi fiber, from the side, to a focused, linearly polarized, laser spot with its linear polarization direction at 45° to the eigenmode axes (Figure 7.9). The hi-bi fiber used was elliptically cored D-fiber and was positioned on an optical flat at the appropriate angle, as shown in Figure 7.8. By axial translation of the fiber, the polarized spot was focused successively at points one birefringence beat length (for a prescribed wavelength) apart. The effect of this was to write an additional birefringence into the fiber, its axes being at 45° to the existing intrinsic axes.

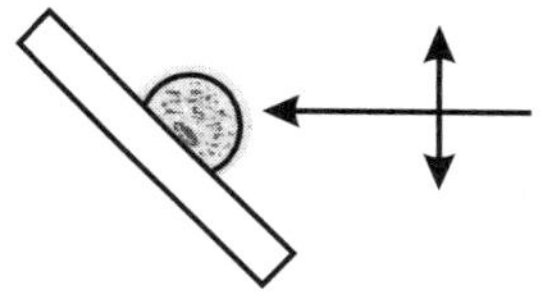

Figure 7.8 Writing a polarization rocking filter.

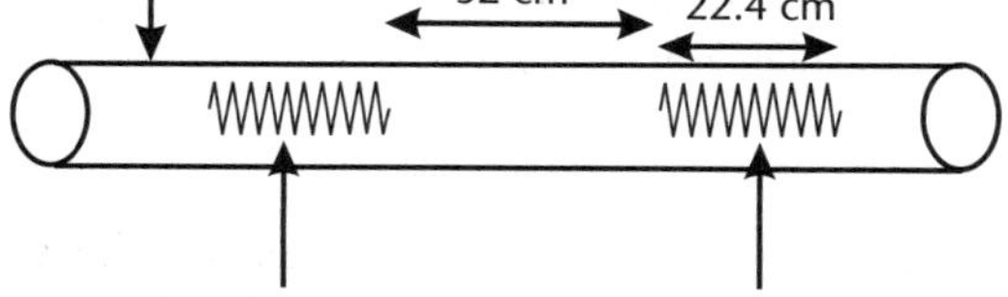

Figure 7.9 Two series-written polarization rocking filters in elliptical-core hi-bi fiber. (*From:* [2]. © 1993 SPIE. Reprinted with permission.)

As a consequence, the resultant (of the two birefringences) axes rock to one side once per beat length.

The effect of this is to couple light, propagating in one of the eigenmodes, into the other, when it encounters the *rocking filter*. In one experiment [2], two rocking filters operating at a wavelength of 787 nm were written externally in an elliptical core D-type fiber (Figure 7.9). The couplers were written with 266-nm light, which has been shown to be capable of introducing a smaller photo-induced absorption than 240-nm light.

The couplers are operated by exciting only one of the fiber polarization eigenmodes (e_x) and monitoring the light in the orthogonal mode (e_y). The coupling efficiency may readily be derived from coupled-mode theory [5].

Experiments with photo-induced couplers written using 266-nm light have shown that these are not erased when heated up to temperatures of 200°C.

Figure 7.10 shows the effect of heating one of the couplers while keeping the other at room temperature. The unequal amplitude of the peak coupling response for the two gratings that appears in the figure is due to the non-uniform response of the spectral analysis system. The broadening of the spectral coupling response of the heated coupler occurred because this coupler was not

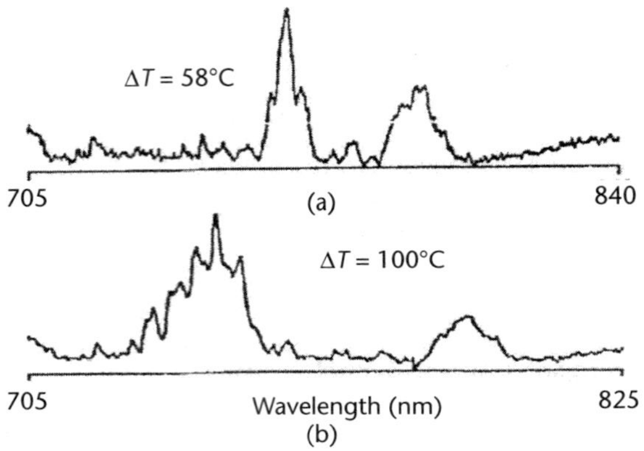

Figure 7.10 (a, b) Shift of coupling wavelength with temperature change. (*From:* [2]. © 1993 SPIE. Reprinted with permission.)

heated uniformly throughout its length. The coupler was heated by laying it on a large hot-plate. This figure shows a resonant wavelength shift of 0.5 nm/°C at 787 nm.

The change, $\delta\lambda$, in the resonant wavelength, λ_0, of the filters for a temperature change δT can be described by the following relation:

$$\delta\lambda/\lambda_0 = (1/\delta n \cdot \partial\delta n/\partial T + 1/\Lambda \cdot \partial\Lambda/\partial T)\,\delta T$$

where the second term in the above relation represents the thermal expansion coefficient, which for silica is $5.5 \times 10^{-7}/°C$.

Suppose now that a conventional photo-generated Bragg grating is overlaid on to a rocking filter of the type just described, and is written into just one of the two eigenmodes, say *Ox*. If broad-spectrum linearly polarized light is launched into the *Ox* eigenmode, it will reflect a particular wavelength, and the dependence of this wavelength on strain σ and temperature T can be written:

$$\delta\lambda_r/\lambda_r = \alpha_1\sigma + \alpha_2 T \tag{7.7a}$$

whereas the resonant coupling (for e_x into e_y) wavelength for the rocking filter will be

$$\delta\lambda_c/\lambda_c = \alpha_3\sigma + \alpha_4 T \tag{7.7b}$$

where all the α_n are different and known. Clearly, equations (7.7) allow σ and T to be determined independently. A measurement of λ_r can be made conveniently by observing the transmission notch (using a broadband source) in one eigenmode, while λ_c can be correspondingly measured by observing the coupling peak (using the same source) in the other eigenmode.

The values of α_1 and α_2 have been measured as [6, 7]

$$\alpha_1 = +0.74$$

$$\alpha_2 = +8.2 \cdot 10^{-6}\ K^{-1}$$

$$\alpha_3 = -1.87$$

$$\alpha_4 = -6.35 \cdot 10^{-4}\ K^{-1}$$

A multiplexing of such overlaid photo-generated grating pairs thus comprises, potentially, an effective quasi-distributed sensor arrangement for the simultaneous, and independent, measurement of strain and temperature.

7.4 Conclusions

In this chapter we have seen how nonlinear effects in optical fibers can give rise to disadvantages, but can also be used to advantage in optical systems.

The effects themselves stem from the fact that the amplitude of the electric field of the propagating optical wave rivals that of the atoms in the material that comprises the optical medium, and thus disturbs it. Because this field is intrinsically directional, so also are the resulting nonlinear effects, and thus it is unsurprising that there are consequences in regard to polarization behavior.

In science, there is nothing to be feared once proper understanding has been acquired (politics aside!).

References

[1] Hill, K. O., et al., "Photo-Sensitivity in Optical-Fiber Waveguides," *Appl. Phys. Lett.*, Vol. 31, pp. 647–652.

[2] Rogers, A. J., and V. A. Handerek, "Static and Dynamic Fiber Polarization Grating Couplers for Sensing Applications," *SPIE's International Symposium on Optical Tools for Manufacturing and Advanced Instrumentation*, Boston, MA, Vol. 2071, paper 2071-06, 1993, pp. 49–58.

[3] Handerek, V. A., A. J. Rogers, and I. Cokgor, "Detection of Localized Polarization Mode Coupling in Polarization-maintaining Fibers," *Proc. of the 8th International Conference on Optical Fiber Sensors (OFS 8)*, Monterey, CA, 1992, pp. 250–253.

[4] Meltz, G., W. W. Morey, and W. H. Glen, "Formation of Bragg Gratings in Optical Fibers by a Transverse Holographic Method," *Opt. Lett.*, Vol. 14, 1984, pp. 823–825.

[5] Yariv, A., and P. Yeh, *Optical Waves in Crystals*, New York: John Wiley & Sons, Chapter 6, p. 155.

[6] Morey, W. W., J. R. Dunphy, and G. Meltz, "Multiplexing Fiber Grating Sensors," *SPIE Proceedings*, Vol. 1586, 1984, 1991, pp. 216–224.

[7] Huang, S. Y., J. N. Blake, and B. Y. Kim, "Perturbation Effects in Mode Propagation in Highly Elliptical Core Two-Mode Fibers," *Journal of Lightwave Technology*, Vol. 8, No. 23, 1990.

8

Epilogue

Optical fibers, together with their many, wide-ranging photonic accessories, have the potential to transform society. Their potential lies primarily in their ability to convey information, at unprecedentedly high rates, for both business and domestic purposes.

The polarization properties of optical fibers have hitherto been less well studied than has their ability to transfer information via the propagation of raw optical power, but these properties are now becoming much more important. There are two reasons for this.

First, polarization properties relate to either intrinsic or imposed asymmetries in the fiber structure and, in telecommunications applications which use modern, high-quality, monomode optical fibers, these asymmetries do not become important until we seek to use bit rates in excess of ~40 Gbps, rates which are only just now being contemplated. At these bit rates, the small differences in the propagation velocities of the various polarization states lead to differences in bit-arrival time, which become significant in relation to the bit period and thus limit the communications bandwidth. The resulting problem of PMD has to be addressed, and the fiber's polarization properties have to be well understood in order to do this.

Second, in the application of optical fibers to information gathering with sensors, the consequences of these asymmetries comprise an advantage rather than a problem. In this domain we can allow a parameter, whose value we need to sense, to impose an asymmetry on the fiber. As a result, a sensing of the polarization properties allows a corresponding sensing of the parameter; and this can be done very sensitively, because a polarimetric sensor is essentially a "phase-disturbance" device, providing the same kind of sensitivity as is available with optical interferometric systems; these are well recognized as among the most sensitive of all. These features, in conjunction with the linear, flexible, one-dimensional nature of the optical fiber, allow for a valuable range of distributed, integrating (including line-averaging), and tailorable point-sensing systems and devices, all of which are now being used increasingly in industrial measurement.

For the future, we can look towards a convergence of these two platforms (telecommunications and measurement). Information gathered by distributed polarimetric sensors can be communicated rapidly and efficiently by fibers to central control points of processing, diagnostic, and decision-making capability (perhaps employing parallel-processing optical or quantum computation), followed by consequent actions communicated over the same pathways. Present-day developments involving new types of optical fibers, such as photonic-crystal fibers (PCF), for example, and new photonic materials resulting from advances in nanotechnology, could well allow a new range of optical features, linear and nonlinear, to be harnessed to these tasks.

By such means, optical fibers and photonics can, in the foreseeable future, provide an information superstructure that can form the basis of a society much better able to benefit from the present rapid pace of technological advance. We may look forward with confidence!

Appendix A: Maxwell's Equations

Maxwell's equations may be expressed in the vectorial form:

1. div $\mathbf{D} = \rho$ (Gauss' theorem)

2. div $\mathbf{B} = 0$ (no free magnetic poles)

3. curl $\mathbf{E} = -\dfrac{\partial \mathbf{B}}{\partial t}$ (Faraday's law of induction + Lenz's law)

4. curl $\mathbf{H} = \dfrac{\partial \mathbf{D}}{\partial t} + \mathbf{j}$ (Ampere's circuital theorem + Maxwell's displacement current)

where ρ is the density of electric charge, $\mathbf{j}$ is the current density, $\mathbf{B} = \mu\mathbf{H}$, and $\mathbf{D} = \epsilon\mathbf{E}$.

In free space, $\mu = \mu_0$; $\epsilon = \epsilon_0$, $\rho = 0$ and $\mathbf{j} = 0$, so that these equations become

1. div $\mathbf{E} = 0$

2. div $\mathbf{H} = 0$

3. curl $\mathbf{E} = -\mu_0 \dfrac{\partial \mathbf{H}}{\partial t}$

4. curl $\mathbf{H} = \epsilon_0 \dfrac{\partial \mathbf{E}}{\partial t}$

Taking the curl of equation (3), we have the mathematical identity

$$\text{curl curl } \mathbf{E} = \text{grad div } \mathbf{E} - \nabla^2 \mathbf{E}$$

so that

$$\text{curl curl } \mathbf{E} = \mu_0 \text{ curl} \frac{\partial \mathbf{H}}{\partial t} = -\mu_0 \frac{\partial}{\partial t} \text{curl } \mathbf{H}$$

and thus, since div $\mathbf{E} = 0$,

$$\nabla^2 \mathbf{E} = \epsilon_0 \mu_0 \frac{\partial^2 \mathbf{E}}{\partial t^2} \tag{A.1}$$

This is a wave equation for $\mathbf{E}$ with wave velocity:

$$c_0 = \frac{1}{(\epsilon_0 \mu_0)^{1/2}}$$

There will clearly be a similar solution for $\mathbf{H}$, from symmetry.

A sinusoidal solution for $\mathbf{E}$ is

$$E_x = E_0 \exp[i(\omega t - kz)]$$

In this case, we have from (A.3), with the resolution $\mathbf{E} = E_x \mathbf{i} \_ + E_y \mathbf{j} + E_z \mathbf{k}$:

$$\text{curl } \mathbf{E} = \begin{vmatrix} \mathbf{i} & \mathbf{j} & \mathbf{k} \\ \frac{\partial}{\partial x} & \frac{\partial}{\partial y} & \frac{\partial}{\partial z} \\ E_x & E_y & E_z \end{vmatrix} = \mathbf{j} \frac{\partial E_x}{\partial z} = -\mu_0 \frac{\partial \mathbf{H}}{\partial t} \tag{A.2}$$

(since $E_y = E_z = 0$; $\partial E_x / \partial y = 0$).

Thus $\mathbf{H}$ can have only a y component ($\mathbf{j}$ vector), and we have

$$H_y = H_0 \exp[i(\omega t - kz)]$$

as the corresponding value for H_y.

Moreover, using (A.2),

$$\frac{E_0}{H_0} = \left(\frac{\mu_0}{\epsilon_0}\right)^{1/2} = Z_0$$

Z_0 is called the electromagnetic impedance, of free space in this case.

Quite generally:

$$\frac{|\mathbf{E}|}{|\mathbf{H}|} = \left(\frac{\mu}{\epsilon}\right)^{1/2} = Z \qquad \text{(A.3)}$$

The energy stored per unit volume in an electromagnetic wave is given (from elementary electomagnetics) by

$$U = \frac{1}{2}(\mathbf{D} \cdot \mathbf{E} + \mathbf{B} \cdot \mathbf{H}) = \frac{1}{2}(\epsilon \mathbf{E}^2 + \mu \mathbf{H}^2)$$

From (A.3) we have

$$U = \epsilon \mathbf{E}^2 = \mu \mathbf{H}^2$$

so that the energy stored in each of the two fields is the same. The energy crossing unit area per second in the Oz direction for components E_x and H_y will be

$$cU = \frac{1}{(\epsilon\mu)^{1/2}} \frac{1}{2}\left(\epsilon E_x^2 + \mu H_y^2\right)$$

$$= \left(\frac{\epsilon}{\mu}\right)^{1/2} E_x^2 = \left(\frac{\mu}{\epsilon}\right)^{1/2} H_y^2 = E_x H_y = \mathbf{E} \times \mathbf{H}$$

This quantity, the vector product of **E** and **H**, is the Poynting vector; that is,

$$\Pi = \mathbf{E} \times \mathbf{H}$$

and represents the flux of energy through unit area in the direction of wave propagation. Its mean value over one cycle of the optical wave therefore represents the mean power per unit area in the optical propagation and is thus equal to the intensity (or irradiance) of the wave.

Appendix B: The Fourier Inversion Theorem

The Fourier inversion theorem states: if $A(\alpha)$ is the Fourier transform (FT) of $f(x)$, then $f(x)$ is the inverse FT of $A(\alpha)$.

The proof is straightforward. If

$$A(\alpha) = \int_{-\infty}^{\infty} f(x) \exp(-i\alpha x)\, dx$$

then the inverse FT of $A(\alpha)$ is

$$\int_{-\infty}^{\infty} A(\alpha) \exp(i\alpha x')\, d\alpha = \int_{-\infty}^{\infty}\int_{-\infty}^{\infty} f(x) \exp(-i\alpha x)\, dx \exp(i\alpha x')\, d\alpha$$

$$= \int_{-\infty}^{\infty}\int_{-\infty}^{\infty} f(x) \exp[-i\alpha(x - x')]\, dx\, d\alpha$$

Integrating with respect to α, we obtain

$$\int_{-\infty}^{\infty} f(x) \left\{ \frac{\exp[-i\alpha(x - x')]}{-i(x - x')} \right\}_{\alpha=-\infty}^{\alpha=\infty} dx$$

Now the function within the square brackets can be written as

$$\lim_{\alpha \to \infty} \left[\frac{2 \sin \alpha(x - x')}{x - x'} \right]$$

and this is clearly the Dirac δ-function: $2\pi\delta(x \_ - x')$.

Hence,

$$\int_{-\infty}^{\infty} A(\alpha) \exp(i\alpha x')\, d\alpha = 2\pi \int_{-\infty}^{\infty} \delta(x - x')\, f(x)\, dx$$

By definition, the δ-function is nonzero only at $x = x'$; hence,

$$\int_{-\infty}^{\infty} A(x) \exp(i\alpha x')\, dx = 2\pi f(x')$$

and the proposition is proved apart from the factor of 2π. It is for this reason that the two sides are often divided by $\sqrt{2\pi}$, so that

$$A'(\alpha) = \frac{A(\alpha)}{\sqrt{2\pi}}$$

$$f'(x) = \sqrt{2\pi}\, f(x)$$

and then

$$A'(\alpha) = \frac{1}{\sqrt{2\pi}} \int_{-\infty}^{\infty} f(x) \exp(-i\alpha x)\, dx$$

$$f'(x) = \frac{1}{\sqrt{2\pi}} \int_{-\infty}^{\infty} A(\alpha) \exp(i\alpha x)\, d\alpha$$

so each is now a true inverse transform of the other. This relationship is often expressed by use of the notation

$$A'(\alpha) = f(x)$$

Appendix C: The Polarization Ellipse

When referred to rectangular Cartesian axes Ox, Oy, the two electric field components of any polarized optical wave may be written:

$$E_x = e_x \cos(\omega t - kz + \delta_x) \tag{C.1}$$

$$E_y = e_y \cos(\omega t - kz + \delta_y)$$

It is straightforward to eliminate $(\omega t - kz) = \tau$, say, from these equations, as follows:

$$\frac{E_x}{e_x} = \cos \tau \cos \delta_x - \sin \tau \sin \delta_x$$

$$\frac{E_y}{e_y} = \cos \tau \cos \delta_y - \sin \tau \sin \delta_y$$

so that

$$\frac{E_x}{e_x} \sin \delta_y - \frac{E_y}{e_y} \sin \delta_x = \cos \tau \sin \delta$$

$$\frac{E_x}{e_x} \cos \delta_y - \frac{E_y}{e_y} \cos \delta_x = \sin \tau \sin \delta$$

where $\delta = \delta_y - \delta_x$. Squaring and adding these gives

$$\left(\frac{E_x}{e_x}\right)^2 + \left(\frac{E_y}{e_y}\right)^2 - 2\frac{E_x E_y}{e_x e_y}\cos\delta = \sin^2\delta \tag{C.2}$$

which is the polarization ellipse referred to E_x, E_y (see Figure 3.2).

To find the ellipticity and orientation of this ellipse, we may cast it into the standard form,

$$\frac{x^2}{a^2} + \frac{y^2}{b^2} = 1$$

by a rotation of the axes through an angle α. The new field components E'_x, E'_y are related to E_x, E_y by

$$E_x = E'_x \cos\alpha - E'_y \sin\alpha$$
$$E_y = E'_x \sin\alpha + E'_y \cos\alpha$$

Substituting these into (C.2), we have

$$\begin{aligned} &E'^2_x\left(\frac{\cos^2\alpha}{e_x^2} + \frac{\sin^2\alpha}{e_y^2} - \frac{\sin 2\alpha\cos\delta}{e_x e_y}\right) \\ &+ E'^2_y\left(\frac{\sin^2\alpha}{e_x^2} + \frac{\cos^2\alpha}{e_y^2} + \frac{\sin^2\alpha\cos\delta}{e_x e_y}\right) \\ &- E'_x E'_y\left(\frac{\sin 2\alpha}{e_x^2} - \frac{\sin 2\alpha}{e_y^2} + \frac{2\cos^2\alpha\cos\delta}{e_x e_y} - \frac{2\sin^2\alpha\cos\delta}{e_x e_y}\right) = \sin^2\delta \end{aligned} \tag{C.3}$$

Now to cast this into the required standard form, the coefficient of the cross-product term $E'_x E'_y$ is equated to zero, giving the value of α as

$$\tan 2\alpha = \frac{2e_x e_y \cos\delta}{\left(e_x^2 - e_y^2\right)}$$

If we now define an angle β such that

$$\frac{e_y}{e_x} = \tan\beta$$

then

$$\tan 2\alpha = \tan 2\beta \cos \delta \tag{C.4}$$

Substituting this into (C.3) and defining a new angle χ such that

$$\tan \chi = \pm \frac{b}{a}$$

we find that

$$\sin 2\chi = -\sin 2\beta \sin \delta \tag{C.5}$$

Hence the orientation (β) and the ellipticity b/a of the ellipse are now determinable from the earlier parameters e_x, e_y, and δ.

Taking now the original axes E_x, E_y, arbitrarily chosen for measurement of the Stokes parameters,

$$\begin{aligned} S_0 &= I(0°, 0) + I(90°, 0) \\ S_1 &= I(0°, 0) + I(90°, 0) \\ S_2 &= I(45°, 0) - I(135°, 0) \\ S_3 &= I(45°, \pi/2) - I(135°, \pi/2) \end{aligned}$$

where, as described in Section 3.8, $I(\vartheta, \epsilon)$ denotes the intensity of the incident light passed by a linear polarizer set at angle ϑ to E_x, after the E_y component has been retarded by angle ϵ as a result of the insertion (or not) of a quarter-wave plate with its axes parallel with E_x, E_y [see Figure 3.14(a)]. Using the original expressions for E_x, E_y from (C.1), it is clear that

$$\begin{aligned} S_0 &= |E_x|^2 + |E_y|^2 = e_x^2 + e_y^2 \\ S_1 &= |E_x|^2 - |E_y|^2 = e_x^2 - e_y^2 \\ S_2 &= \frac{1}{2}|E_x + E_y|^2 - \frac{1}{2}|E_x - E_y|^2 = 2e_x e_y \cos \delta \\ S_3 &= \frac{1}{2}|E_x + iE_y|^2 - \frac{1}{2}|E_x - iE_y|^2 = 2e_x e_y \sin \delta \end{aligned}$$

From (C.4) and (C.5) it now follows that

$$\frac{S_3}{S_0} = \sin 2\chi$$

and

$$\frac{S_2}{S_1} = \tan 2\alpha$$

with, also,

$$S_0^2 = S_1^2 + S_2^2 + S_3^2$$

Hence the measurement of the Stokes parameters provides a quick and convenient method for complete specification of the polarization ellipse.

The degree of polarization, in the case of partially polarized light, is given by

$$\eta = \frac{S_1^2 + S_2^2 + S_3^2}{S_0^2}$$

Appendix D: Elliptical Birefringence

For an element possessing both linear birefringence, δ, and circular birefringence, 2ρ, we can write the Jones matrix, referred to the linear birefringence axes OX (fast) and OY (slow) as

$$\mathbf{M} = \begin{pmatrix} \alpha + i\beta & -\gamma \\ \gamma & \alpha - i\beta \end{pmatrix}$$

where

$$\alpha = \cos \Delta$$

$$\beta = \delta/2 \cdot \sin \Delta/\Delta$$

$$\gamma = \rho \sin \Delta/\Delta$$

$$\Delta = (\delta^2/4 + \rho^2)$$

The eigenvectors $\begin{pmatrix} X \\ Y \end{pmatrix}$ and eigenvalues, λ, of $\mathbf{M}$ are the solutions of the eigenvalue equation:

$$\mathbf{M}\begin{pmatrix} X \\ Y \end{pmatrix} = \lambda \begin{pmatrix} X \\ Y \end{pmatrix}$$

The eigenvalues solutions are

$$\lambda_1 = \alpha + (\alpha^2 - 1)^{1/2}$$

$$\lambda_2 = \alpha - (\alpha^2 - 1)^{1/2}$$

Substituting for α, we find

$$\lambda_1 = \exp(i\Delta)$$

$$\lambda_2 = \exp(-i\Delta)$$

Thus λ_1 represents a phase advance of Δ, and λ_2 a phase delay of Δ.

The corresponding eigenvectors are

$$\begin{pmatrix} X_1 \\ Y_1 \end{pmatrix} = K_1 \begin{pmatrix} \gamma \\ i\beta - (\alpha^2 - 1)^{1/2} \end{pmatrix}$$

$$\begin{pmatrix} X_2 \\ Y_2 \end{pmatrix} = K_2 \begin{pmatrix} \gamma \\ i\beta + (\alpha^2 - 1)^{1/2} \end{pmatrix}$$

where K_1 and K_2 are arbitrary constants.

Now with

$$\alpha = \cos \Delta$$

we may write

$$(\alpha^2 - 1)^{1/2} = i \sin \Delta$$

and thus Y_1 and Y_2 are purely imaginary. Hence, the X and Y components, for both eigenvectors, are orthogonal and $\pi/2$ out of phase. Hence, the eigenvectors represent ellipses with the ellipse axes parallel with the axes OX and OY for, in an ellipse formed from two orthogonal sinusoids at the same frequency, only the components along the axes of the ellipse are $\pi/2$ out of phase. That is,

$$x = a \sin \omega t$$

$$y = b \cos \omega t$$

hence,

$$x^2/a^2 + y^2/b^2 = 1$$

which is the equation for an ellipse with axes of length a and b parallel with Ox and Oy, respectively. It follows that the axes of the ellipses are parallel with the axes of linear birefringence.

The values of the ellipticities are the ratios of the minor to the major axes and hence are given by

$$e_1 = (\beta - \sin \Delta)/\gamma$$
$$e_2 = \gamma/(\beta + \sin \Delta)$$

Substituting for β and γ we have

$$e_1 = (\delta - 2\Delta)/2\rho$$
$$e_2 = 2\rho/(\delta + 2\Delta)$$

Now

$$(\delta - 2\Delta) = \delta - (\delta^2 + 4\rho^2)^{1/2}$$

and hence this must be negative. It follows that e_1 is negative whilst e_2 is positive. This means that the two ellipses are circumscribed in opposite directions.

Comparing the actual magnitudes of the ellipticities, we note that these are equal, for

$$(2\Delta - \delta)/2\rho = 2\rho/(2\Delta + \delta)$$

because

$$4\rho^2 = 4\Delta^2 - \delta^2$$

To summarize, the orthogonal ellipses possess the same ellipticities and are circumscribed in opposite directions. Their relative propagation phase delay is

$$2\Delta = (\delta^2 + 4\rho^2)^{1/2}$$

Finally, if

$$\tan 2\chi = 2\rho/\delta$$

we see that

$$e_1 = e_2 = e(say) = (2\Delta - \delta)/2\rho = (\delta^2/4\rho^2 + 1)^{1/2} - \delta/2\rho$$

Hence,

$$e = (\cot^2 2\chi + 1)^{1/2} - \cot 2\chi = \cos ec2\chi - \cot 2\chi$$

that is,

$$e = \tan \chi$$

Appendix E: Second Harmonic Generation

The generation of the second harmonic from a fundamental optical wave, when it is passing through a nonlinear optical material, must properly be treated with Maxwell's equations as the starting point.

For an insulating medium of permittivity ϵ and permeability μ, we can write Maxwell's equations in the form:

$$1.\ \text{div}\ \mathbf{E} = 0$$

$$2.\ \text{div}\ \mathbf{H} = 0$$

$$3.\ \text{curl}\ \mathbf{E} = -\mu\mu_0 \frac{\partial \mathbf{H}}{\partial t} = -\frac{\partial \mathbf{B}}{\partial t}$$

$$4.\ \text{curl}\ \mathbf{H} = \epsilon\epsilon_0 \frac{\partial \mathbf{E}}{\partial t} = \frac{\partial \mathbf{D}}{\partial t}$$

Now assuming that the nonlinear electric polarization vector lies in the same direction as the applied electric field, the first two terms of its expansion can be written in the form:

$$\mathbf{P} = \epsilon\chi_1\mathbf{E} + \epsilon_0\chi_2|\mathbf{E}|\mathbf{E}$$

Hence we have

$$\mathbf{D} = \epsilon_0 \mathbf{E} + \mathbf{P} = \epsilon_0 \mathbf{E} + \epsilon_0 \chi_1 \mathbf{E} + \epsilon_0 \chi_2 |\mathbf{E}| \mathbf{E}$$

or

$$\mathbf{D} = \epsilon \epsilon_0 \mathbf{E} + \epsilon_0 \chi_2 |\mathbf{E}| \mathbf{E}$$

(since $\epsilon = 1 + \chi$). Substituting this expression for $\mathbf{D}$ in Maxwell's equation 4, we have

$$\text{curl } \mathbf{H} = \epsilon \epsilon_0 \frac{\partial \mathbf{E}}{\partial t} + \epsilon_0 \chi_2 \frac{\partial \left(|\mathbf{E}| \mathbf{E} \right)}{\partial t} \tag{E.1}$$

Taking now the curl of Maxwell's equation 3, and using the mathematical identity

$$\text{curl curl } \mathbf{E} = \text{grad div } \mathbf{E} - \nabla^2 \mathbf{E}$$

we find that

$$\nabla^2 \mathbf{E} = \mu \mu_0 \frac{\partial (\text{curl } \mathbf{H})}{\partial t}$$

(since div $\mathbf{E} = 0$). Using (E.1), we now have that

$$\nabla^2 \mathbf{E} = \mu \mu_0 \epsilon \epsilon_0 \frac{\partial^2 \mathbf{E}}{\partial t^2} + \mu \mu_0 \epsilon \epsilon_0 \chi_2 \frac{\partial^2 \left(|\mathbf{E}| \mathbf{E} \right)}{\partial t^2} \tag{E.2}$$

Suppose now that we consider a solution for this equation, which consists of the fundamental and a second harmonic, of the form:

$$\mathbf{E}(z, t) = \mathbf{E}_1(z) \exp[i(\omega_1 t - k_1 z)] + \mathbf{E}_2(z) \exp[i(2\omega_1 - k_2 z)]$$

This can now be substituted into (E.2) in order to determine the relationship between the first and second harmonic components. Essentially this relationship will tell us how the second harmonic is "generated" from the fundamental. It is clear that this generation can only result from the square of the fundamental component since only that leads to the correct frequency. It is therefore necessary only to deal with terms which oscillate with frequency $2\omega_1$; all other terms will operate independently.

We assume that all vectors again are parallel, and orthogonal to the propagation direction Oz.

We shall deal with each side of (E.2) in turn. Remembering that we are only concerned with terms in $2\omega_1 t$, the right-hand side becomes, on substituting $E(z, t)$,

$$(\nabla^2 E) = -\mu\mu_0 \epsilon\epsilon_0 4\omega_1^2 E_2(z) \exp[i(2\omega_1 t - k_2 z)]$$
$$- \mu\mu_0 \epsilon_0 \chi_2 4\omega_1^2 E_1^2(z) \exp[i(2\omega_1 t - k_1 z)]$$

where μ is assumed constant for both components. The left-hand side becomes

$$\nabla^2 E(z, t) = \frac{\partial^2}{\partial z^2} E_2(z) \exp[i(2\omega_1 t - k_2 z)]$$
$$- \left[k_2^2 E_2(z) + ik_2 \frac{\partial E_2(z)}{\partial z} \right] \exp[i(2\omega_1 t - k_2 z)]$$

where, for this last expression, it has been assumed that

$$\left| 2k_2 \frac{\partial E_2(z)}{\partial z} \right| \gg \left| \frac{\partial^2 E_2(z)}{\partial z^2} \right|$$

that is, that $\partial E_2(z)/\partial z$ is sensibly constant over one second harmonic wavelength. Equating the right-hand side and left-hand side of (E.2) and canceling the common factor $(\exp(2i\omega_1 t))$, we obtain

$$4\mu\mu_0 \epsilon_2 \epsilon_0 \omega_1^2 E_2(z) \exp(-ik_2 z) + 4\mu\mu_0 \epsilon_0 \chi_2 \omega_1^2 E_1^2(z) \exp(-2ik_1 z)$$
$$= k_2^2 E_2(z) \exp[-ik_2 z] + ik_2 \frac{\partial E_2(z)}{\partial z} \exp(-ik_2 z)$$

Now we know that

$$c_2^2 = \frac{1}{\mu\mu_0 \epsilon_2 \epsilon_0}; \quad \frac{2\omega_1}{k_2} = c_2$$

(since the permittivity and permeability constants will refer to the propagation for which they are the coefficients, in this case the second harmonic).

It follows that the first terms on each side cancel out to give

$$ik_2 \frac{\partial E_2(z)}{\partial z} \exp(-ik_2 z) = 4\mu\mu_0 \epsilon_0 \chi_2 \omega_1^2 E_1^2(z) \exp(-2ik_1 z)$$

or

$$\frac{\partial E_2(z)}{\partial z} = \frac{-2i\chi_2\omega_1}{c_2\epsilon_2} E_1^2(z) \exp[-i(2k_1 - k_2)z] \tag{E.3}$$

This is the "generator" equation, showing the relationship between the spatial growth of $E_2(z)$ as a result of $E_1^2(z)$. In order to determine the value of $E_2(z)$ after a length L of generation in the nonlinear crystal, (E.3) must be integrated:

$$E_2(L) = \frac{-2i\chi_2\omega_1}{c_2\epsilon_2} E_1^2(L) \frac{\exp[-i(2k_1 - k_2)L] - 1}{-i(2k_1 - k_2)}$$

Hence the intensity of $E_2(L)$ will be proportional to

$$E_2(L)E_2^*(L) = \frac{4\chi_2^2\omega_1^2 L^2 E_1^4(L)}{c_2^2\epsilon_2^2} \left[\frac{\sin\left(k_1 - \frac{1}{2}k_2\right)L}{\left(k_1 - \frac{1}{2}k_2\right)L} \right]^2$$

This is seen to have the same form as (4.7), which was derived from (largely) physical intuition. It is instructive to compare the two equations in detail, and this will be left as an exercise for the reader.

About the Author

Alan Rogers is presently a visiting professor of electronics at the University of Surrey. Previously he was the head of the Department of Electronic Engineering at King's College London. He has published more than 200 papers in the area of photonics and has initiated 14 patents. Professor Rogers holds a double first in natural sciences from the University of Cambridge and a Ph.D. in space physics from University College London. He is a Fellow of the Institute of Physics, a Fellow of the Institution of Engineering and Technology, and a Senior Member of the Institute of Electronic and Electrical Engineers.

Index

Recent Related Artech House Titles

New Photonics Technologies for the Information Age: The Dream of Ubiquitous Services, Shoichi Sudo, and Katsunari Okamoto, editors

Optical Document Security, Third Edition, Rudolph L. van Renesse

Optical Transmission Systems Engineering, Milorad Cvijetic

Optics of Quantum Dots and Wires, Garnett W. Bryant and Glenn S. Solomon

Semiconductor Nanostructures for Optoelectronics Applications, Todd Steiner, editor

Understanding Optical Fiber Communications, Alan J. Rogers